ORGANOCHLORINE

Edited by **Aurel Nuro**

Organochlorine

http://dx.doi.org/10.5772/intechopen.74499

Edited by Aurel Nuro

Contributors

Mekonnen Maschal Tarekegn, Efrem Sisay Akele, Yoshiharu Mitoma, Cristian Simion, Yumi Katayama, Fuchong Li, Guangbi Gong, Yansheng Liu, Linlei Wang, Xu Li, Tianqiong Ma, Aurel Nuro

Notice

Statements and opinions expressed in the chapters are these of the individual contributors and not necessarily those of the editors or publisher. No responsibility is accepted for the accuracy of information contained in the published chapters. The publisher assumes no responsibility for any damage or injury to persons or property arising out of the use of any materials, instructions, methods or ideas contained in the book.

First published in London, United Kingdom, 2018 by IntechOpen

IntechOpen is the global imprint of INTECHOPEN LIMITED, registered in England and Wales, registration number: 11086078, The Shard, 25th floor, 32 London Bridge Street

London, SE19SG – United Kingdom

Printed in Croatia

British Library Cataloguing-in-Publication Data

A catalogue record for this book is available from the British Library

Additional hard copies can be obtained from orders@intechopen.com

Organochlorine, Edited by Aurel Nuro

p. cm.

Print ISBN 978-1-78984-263-0

Online ISBN 978-1-78984-264-7

We are IntechOpen,
the world's leading publisher of
Open Access books
Built by scientists, for scientists

3,800+
Open access books available

116,000+
International authors and editors

120M+
Downloads

151
Countries delivered to

Our authors are among the

Top 1%
most cited scientists

12.2%
Contributors from top 500 universities

CLARIVATE ANALYTICS
BOOK
CITATION
INDEX
INDEXED

WEB OF SCIENCE™

Selection of our books indexed in the Book Citation Index
in Web of Science™ Core Collection (BKCI)

Interested in publishing with us?
Contact book.department@intechopen.com

Numbers displayed above are based on latest data collected.
For more information visit www.intechopen.com

Meet the editor

Aurel Nuro was born in Albania. He graduated with a degree in Chemistry (2002) from the Faculty of Natural Sciences, Tirana University. He received a Master of Science (2004) and PhD (2008) from the Tirana University and "Docent" (2010) and Associated Professor (2012) titles. From 2002, he has been working as a Lecturer and Researcher in the Organic Chemistry group, Department of Chemistry, Faculty of Natural Sciences, University of Tirana. His main research areas are: organic chemistry, organic analysis, instrumental analyses, gas chromatography techniques, environment pollution, food control, ecotoxicology, pesticides, PCB, etc. He has been a coordinator, expert and participant in several national and international projects. More than 160 Master theses in environmental chemistry, food chemistry, pharmacy, etc. were led by him. His publications include articles, books, conference proceedings (more than 150), and he is cited in scientific literature.

Contents

Preface IX

Chapter 1   **Introductory Chapter: Organochlorine 1**
Aurel Nuro

Chapter 2   **Dioxin and Furan Emissions and Its Management Practices 7**
Mekonnen Maschal Tarekegn and Efrem Sisay Akele

Chapter 3   **Mechanistic Considerations on the Hydrodechlorination
Process of Polychloroarenes 25**
Yoshiharu Mitoma, Yumi Katayama and Cristian Simion

Chapter 4   **Application of Heterogeneous Catalysts in Dechlorination of
Chlorophenols 47**
Fuchong Li, Yansheng Liu, Linlei Wang, Xu Li, Tianqiong Ma and
Guangbi Gong

Chapter 1

Introductory Chapter: Organochlorine

Aurel Nuro

Additional information is available at the end of the chapter

http://dx.doi.org/10.5772/intechopen.81271

1. Introduction

Organochlorines (OCs) are organic molecules with chlorine in their structure. There are known a large number of organochlorine compounds. A large amount of chlorinated organic compounds are produced for industrial, agricultural, pharmaceutical, household purposes, etc. In many studies, the main focus is on OC which has been evaluated as environmental contaminants with toxic effects on humans. Different types of organochlorine have been produced throughout the world. Some of the most popular classes are organochlorine pesticides (OCPs), polychlorinated biphenyls (PCBs), dioxins, chlorobenzenes, chlorophenols, chlorinated alkanes, etc. Organochlorine compounds usually have a large molecular mass. They are very stable. Generally, they are molecules of moderate polarity (low solubility in water). This makes OC easily soluble in fats. They were found in almost all environments: air, water, soil, sediments, and biota samples. They can spread out easily in different geographic altitudes and latitudes. Volatile and semi-volatile OCs have the ability to spread far away from the place where they were used. Some studies have reported some organochlorines in the North Pole at the same levels as the areas where they were produced or applied. They have the ability to bioaccumulate easily in biota. Passing through the food chain levels, they increase their concentrations (biomagnifying). Contaminated foods with OCs and exposures to them are their main ways to arrive in the human body. Generally, they display their effects after a relatively long period of exposure. This is the main reason why they are produced and used for a long time before their production and use were banned. The most important health effects that organochlorines can cause are mutagenic, endocrine-disruptor, carcinogenic and central nervous or peripheral disorders. After identification of the consequences for many organochlorines, their production and use in many countries were banned but unfortunately their effects were shown for many other years.

2. Organochlorine pesticides

Organochlorine pesticides were used widely for agricultural purposes after the Second World War. The insecticidal properties of DDT were discovered firstly. After that many other OCPs were synthesized and used. Many tons of DDT, aldrin, heptachlor, lindane, hexachloroben-zene, toxaphene, and many other pesticides with chlorine in their molecules were produced and used all over the world for many years. They affected significantly the growth of agricul-tural products around 1960, bringing what was called the "Green Revolution." Their use was very effective especially in countries that suffered from many diseases spread by insects such as malaria. In early 1970s, their toxic effects were verified firstly in birds and fishes. After that, OCPs were banned in the USA and Europe. OCPs and their degradation products are found in many ecosystems until now because of their persistence and their bioaccumulation ability. The presence of OCPs was reported in many environmental and food studies. Water irriga-tion and rainfall landslides make them possible to pass over surface and underground waters. So, they can spread far away from areas where they have been applied. Bioaccumulation processes make their presence possible in all food chain [1, 2].

3. Polychlorinated biphenyls

PCBs were produced from 1930 to 1977. PCB mixtures were prepared from chlorination of biphenyls in the presence of various catalysts. There are 209 different congeners depending on the number of chlorine atoms and their positions in the molecule. Usually they were used as mixtures (aroclor, kanechlor, etc.). These mixtures are classified and used according to the percentage of chlorine. They were widely used as insulating and hydraulic fluids in numer-ous industrial processes. They were banned in 1980 because of the possible risks to human health and the environment. Higher toxicity presents non- and mono-ortho-substituted con-geners because of their planarity. They are called dioxin-like congeners of PCBs, classified as endocrine disruptor and possible carcinogen to humans [3, 4]. Many accidents have occurred around the world due to the use of PCB: Kyushu, Japan (1968); Hudson River, USA (1977); Brescia, Italy (1999); etc. These accidents and laboratory in vivo/in vitro data verified their toxic ability to humans [5–8].

4. Dioxins and furans

Polychlorinated dibenzo-p-dioxin (PCDDs) and polychlorinated dibenzofurans (PCDFs) are highly toxic, the most dangerous being 2,3,7,8-tetrachlorodibenzo-p-dioxin (TCDD). Testing study on toxicity of the TCDD is used as a mechanism and reference value for other dioxins [9–11]. PCDD/Fs and PCB dioxin-like were verified in different experiments to have the same mechanism of toxicity via the aryl hydrocarbon receptor (AHR). Dioxins can produced in higher temperature as secondary products, e.g. by increate of urban wastes or many indus-trial processes. Dioxins have no common use. PCDD/F even in low levels can cause serious problems to humans and other organisms because of their high toxicity [12, 13]. Various studies have clarified the implications of their presence at ultra-trace levels like that in Seveso, Italy (1976), Belgian PCB/dioxin incident (1999), etc [14, 15].

5. Chlorobenzenes, chlorophenols, and other derivatives

Derivatives with chlorine of benzene, phenol, aniline, nitrobenzene, benzoic acid, phenyl acetic acid, and many other similar compounds are usually produced as raw materials for many syntheses in the pharmaceutical industry, plastic production, pesticide production, and many other synthetic compounds. Some of them were used as pesticides, for example, tecnazene (tetrachloronitrobenzene), Kvintozen (pentachloraniline), etc. Chlorinated derivatives can be impure in many synthetic products. They can be obtained by degradation processes of large chlorinated molecules. They were reported as part of the metabolism of large chlorine molecules in different organisms. Most of them are harmful to the environment and living organisms. Their toxicity is different depending on the number and position of chlorine in the molecules.

6. Legislation on organochlorine compounds

Many studies have shown that chlorine compounds have harmful health effects not only for exposed persons but also for the entire population, it was necessary to regulate the equivalence of international legislation on limitations on the production and use of toxic substances. So several agreements were reached, but the most important is the Stockholm Convention (2001) on Persistent Organic Pollutants (POPs). This convention, adopted by most countries, aimed at the elimination or reduction and use of many OCs including OCPs, PCBs, and dioxins. It provided several phases and guidelines for the immediate prohibition of use, production, and reduction of those substances called POPs. Although in most countries this convention became effective soon, OC presence was reported in certain areas as a result of waste disposal, equipment accidents, or their use under false trademark. Their high persistence is an important factor. Control of the OC should be continuous in environmental and food samples due to their wide spread.

7. Organochlorine distribution, stability, and degradation

Although they have been banned many years ago, levels of OC were reported frequently in different studies. They are widespread in both applied environments and in areas far away from their application sites. This is due to high stability, high bioaccumulation capacity, biomagnification, and the ability to spread out of the application site. Generally, these compounds are difficult to degrade. In the soil or sediment environment, the speed of degradation is much smaller. Different degradation mechanisms are known, such as photochemical degradation, thermal degradation, biological degradation, and chemical degradation [7]. Their degradation products are derivatives which in most cases also contain chlorine and exhibit certain toxicity. The study of degradation mechanisms will affect the speeding up of their elimination processes under practical conditions.

8. Analytical techniques on determination of organochlorines

The levels of chlorine compounds in both environmental and food samples were found in very low (trace or ultra-trace) levels from ppm to ppt. For their qualitative and quantitative determination, it is necessary to use different techniques of extraction such as Soxhlet, ultrasonic extraction, SPE, ASE, etc. The samples of clean-up procedures generally were realized in SPE columns with adsorbents that have different polarities. Analytical determinations of organochlorines were recommended to be achieved by gas chromatography techniques especially coupled with mass spectrometry (GC/MS). GC/MS/MS and LC/MS/MS are recommended in many methods for OC analysis in environmental and food samples [16, 17]. In many standard methods (EN, ASTM, etc.), techniques of simultaneous determination of organochlorine compounds for the same type or different types due to their similarity are described.

9. Preface on organochlorine chapters

This compact book has some data on organochlorine compounds and their degradation products. Clarification of degradation processes for OC is important for polluted ecosystems. Reviews on legislations for organochlorine compounds especially on persistent and toxic OC were shown also in this book. Analytical procedures (extraction techniques, clean-up procedures, equipment, etc.) and experimental data for OC analysis are mentioned briefly on book chapters.

Chapter I, "Service Sector-Based Dioxin and Furan Emissions and Management Techniques," describes PCDD/F as one of the important classes of organochlorine contaminants. This chapter presents the importance of study dioxins because they cause people's health problems. Through various studies that authors have used, this review is focused mainly on emission sources and monitoring of these contaminants in some countries in order to conclude to the situation in Ethiopia. PCDD/Fs are part of the POPs because of their persistence and toxicity. Also, in this chapter, are given data on other chlorinated compounds, classified as POPs, such as organochlorine pesticides, PCB, PBB, etc. The connection between PCDD/F and other POPs could have been clearly demonstrated.

Chapter II, "Mechanistic Considerations on the Hydrodechlorination Process of Polychloroarenes," is a review of different hydrodechlorination mechanisms for organochlorine pollutants such as PCDD/F and PCB. These pollutants are persistent for many years because of their structure. Their presence could be for many years not only in environment but also in food chain because of bioaccumulation processes. Authors have considered different studies of hydrodechlorination mechanisms based on different redox mechanisms. These mechanisms of hydrodechlorination are important processes because degradation products usually are organic compounds with lower toxicity. Understanding these reaction mechanisms can lead to their efficient use in practice. The use of these reactions in incinerator filters and contaminated areas can bring a reduction of PCDD/PCDF and PCB pollution.

Chapter III, "Application of Heterogeneous Catalysts in Dechlorination of Chlorophenols," is a review on heterogeneous catalysts for dechlorination of chlorophenols. These compounds are widely distributed especially in waste and surface waters because of household products, urban wastes and other chlorinated compounds such as pesticides. Heterogeneous catalyst can replace the homogeneous catalyst to solve the catalyst recycling problem, especially for the precious metal catalysts. The heterogeneous catalyst costs are more reduced than homogeneous catalyst. Dechlorination of chlorophenols using heterogeneous catalyst could be useful for dechlorination of other chlorinated compounds.

Author details

Aurel Nuro

Address all correspondence to: aurel.nuro@fshn.edu.al

Faculty of Natural Sciences, University of Tirana, Tirana, Albania

References

[1] Stoytcheva M. Pesticides-formulations, effects, fate. In: Nuro A, Marku E, editors. Chapter 9: Organochlorine Pesticides Residues for Some Aquatic Systems in Albania. Rijeka, Croatia: InTech; 2011. ISBN: 978-953-307-532-7

[2] Jokanović M. Impact of pesticides. In: Nuro A, Marku E, editors. Section 2: Pesticides in the Environment-An Overview of Organochlorinated Pesticide Residues in Albania. Case Study: Porto Romano, Adriatic Sea. USA: Academy Publish.org; 2012. pp. 225-239. ISBN: 978-0-9835850-9-1

[3] Erickson MD. Introduction: PCB properties, uses, oçurrence, and regulatory history. In: Robertson LW, Hansen LG, editors. PCBs: Recent Advances in Environmental Toxicology and Health Effects. Lexington, Kentucky: The University Press of Kentucky; 2001. pp. 131-152

[4] Zuçato E, Calvarese S, Mariani G, Mangiapan S, Grasso P, Guzzi A, et al. Level, sources and toxicity of polychlorinated biphenyls in the Italian diet. Chemosphere. 1999;**38**(12):2753-2765

[5] Safe S. Polychlorinated biphenyls (PCBs), dibenzo-p-dioxins (PCDDs), dibenzofurans (PCDFs), and related compounds: Environmental and mechanistic consideration which support the development of toxic equivalency factors (TEFs). Critical Reviews in Toxicology. 1990;**21**(1):51-88

[6]   ATSDR. Toxicological Profile for Polychlorinated Biphenyls (PCBs). US Department of Health and Human Services. Atlanta, Georgia: Public health Service, Agency for Toxic Substances and Disease Registry; 2000

[7]   Abramowicz DA. Aerobic and anaerobic biodegradation of PCBs: A review. Biotechnology. 1990;**10**(3):241-251

[8]   Ahlborg UG, Brouwer A, Fingerhut MA, et al. Impact of polychlorinated dibenzo-p-dioxins, dibenzofurans, and biphenyls on human and environmental health, with special emphasis on application of the toxic equivalency factor concept. European Journal of Pharmacology. 1992;**228**(4):179-199

[9]   Van den Berg M, Birnbaum LS, Bosveld ATC, Brunstrom B, Cook P, Feely M, et al. Toxic equivalency factors (TEFs) for PCBs, PCDDs and PCDFs for humans and wildlife. Environmental Health Perspectives. 1998;**106**:775-792

[10] Papadopoulos A, Vassiliadou I, Costopoulou D, Papanicolaou C, Leondiadis L. Levels of dioxins and dioxin-like PCBs in food samples on the Greek market. Chemosphere. 2004;**57**:413-419

[11] Focant J-F, Pirard C, De Pauw E. Levels of PCDDs, PCDFs and PCBs in Belgian and international fast food samples. Chemosphere. 2004;**54**:137-142

[12] U.S. Environmental Protection Agency Health Assessment Document for 2,3,7,8-tetrachlorodibenzo-p-dioxin (TCDD) and Related Compounds. Prepared by the Office of Health and Environmental Assessment, Office of Research and Development, Washington, DC. External Review Draft. Vol. 3; 1994b. Report No. EPA/600/BP-92/001c

[13] Hirai a Y, Sakai S-i, Watanabe N, Takatsuki H. Congener-specific intake fractions for PCDDs/DFs and Co-PCBs: Modeling and validation. Chemosphere. 2004;**54**:1383-1400

[14] Schepens PJ, Covaci A, Jorens PG, Hens L, Scharpe S, van Larebeke N. Surprising findings following a Belgian food contamination with polychlorobiphenyls and dioxins. Environmental Health Perspectives. 2001;**109**:101-103

[15] Takekuma M, Saito K, Ogawa M, Matumoto R, Kobayashi S. Levels of PCDDs, PCDFs and Co-PCBs in human milk in Saitama, Japan, and epidemiological research. Chemosphere. 2004;**54**:127-135

[16] Petrick G, Schulz DE, Duinker JC. Clean-up of environmental samles for analysis of organochlorine compounds by gas chromatography with electron-capture detection. Journal of Chromatography. 1988;**435**:241-248

[17] Rene van der Hoff G, van Zoonen P. Trace analysis of pesticides by gas chromatogaphy. Journal of Chromatogaphy A. 1999;**843**:301-322

Dioxin and Furan Emissions and Its Management Practices

Mekonnen Maschal Tarekegn and Efrem Sisay Akele

Additional information is available at the end of the chapter

http://dx.doi.org/10.5772/intechopen.80011

Abstract

Many changes like increment of the population and demanded services, expansion of industries, increasing of transportation demand, etc., have increased the emission of dioxin and furan. There was no indicative research conducted on the quantification and management practices of the unintentionally produced persistent organic pollutants like dioxin and Furan. A UNEP model for dioxin- and furan-related POPs management was commonly used to assess the main anthropogenic sources of dioxin and furan. In this book chapter, UNEP toolkit that was developed in 2013 is used to identify and quantify the sector-based emission of dioxin and furan. About nine main groups of anthropogenic POPs sources such as waste incineration, open burning process, ferrous and nonferrous metal production, etc., explicitly discussed in the report were identified. The case study in Addis Ababa showed that all organizations have no awareness about the dioxin and furan emission issues and follow very weak management styles. Finally, the book chapter suggests the reformulation of the national legal management framework, adaptation of best available technology with less POPs footprint, increasing public and stakeholder's awareness and participation and capacitating the concerned government organization.

Keywords: persistence, polychlorinated dibenzo dioxin, polychlorinated dibenzo furans, Stockholm Convention

1. Introduction

POPs are organic compounds that resist chemical, biological, and photolytic degradation due to their inherent characteristics. Their low water solubility and high lipid solubility facilitate their bioaccumulation in fatty tissues of living organisms. Many are also semi-volatile, which enable them to be transported long distances through the atmosphere. Due to its persistence behavior, POPs are today present all over the world, found in every major climatic zone and geographic sector, including deserts, the Arctic, and the Antarctic were no major local POPs sources exist [1].

There exist several different forms of POPs, natural as well as anthropogenic. Those noted for their persistence and ability to bioaccumulate include many of the first-generation organochlorine insecticides, e.g., Dieldrin and DDT, as well as industrial products or by-products such as PCBs and dioxins. Due to their persistence and ability to accumulate and biomagnifies in living tissues, they can cause harm in the environment for an extensive amount of time [1].

In May 2001, the Stockholm Convention on Persistence Organic pollutant is one of a global, legally binding instrument, aimed at protecting human health and environment across the world from the harmful impact of persistent organic pollutant. According to the Stockholm Convention, this convention perhaps best understood as having five essential aims, such as eliminating dangerous pops by starting from the 12 worst, supporting the transition to safer alternatives, targeting additional POPs for action, cleaning up old stockpiles and equipment containing pops, and working together for POPs free nature.

Regarding the convention, Ethiopia has been proclaimed the ratification of this convention on 2nd day July 2002, Proclamation No. 279/2002, which is the Stockholm Convention on Persistence Organic Pollutant. There are articles that stated in the convention for the management of persistent organic pollutant. Article 5 of the convention deals with the unintentionally produced POPs. It requires each party to take measures to reduce the total releases derived from anthropogenic sources of Annex C chemicals, i.e., HCB, PCBs, dioxins, and furans.

Developing an action plan to identify, characterize, and address the unintentional release of these chemicals is the major obligation of each party. The action plan should evaluate current and projected releases, develop source inventories, and release estimates. It should also evaluate the efficacy of laws and policies relating to the management of such releases. In addition to the action plan, each party is required to (i) promote feasible, practical measures that can expeditiously achieve a significant on reduction of these releases; (ii) promote and/or require use of substitute materials or processes to prevent the formation of these chemicals; (iii) promote and implement, in accordance with the action plan, the use of best available techniques and best environmental practices for existing and any newly identified sources of the chemicals.

The main sources of unintentionally produced POPs cover a wide range of economic activities including industrial processes, such as ferrous and nonferrous metals production, cement and other minerals production, and production and use of chemicals and consumer goods, such as manufacture of pulp and paper, chemicals, petroleum, textiles, and leather products. The other categories include waste incineration, power generation and other fuel burning, transport; uncontrolled combustion processes such as agricultural and forest fires, drying of biomass, crematoria, dry cleaning, and tobacco smoking are also considered as having the potential for formation and release of these chemicals to the environment [2].

This book chapter evaluates the sources and management practice of the unintentionally produced persistence organic pollutants such as dioxin and furan especially in the service sectors.

Potentially available literatures covering the concept of persistence organic pollutant, the birth of the Stockholm Convention, identification, and quantification of unintentionally produced persistent organic pollutant, environmental and health impact of persistent organic pollutant, and policy and regulation framework of pops management were discussed. The global experience and practices of POPs are also discussed.

2. Methodology

This chapter was organized by reviewing of related literatures, collecting primary information, and compiling secondary from various sources. Related peer reviewed articles were compiled online from reputable journals, and secondary data were collected from relevant government offices.

A sector source-based cross-sectional study was carried out to generate quantitative information of UPOPs amount and evaluate sectoral offices management practices specific to dioxin and furan in the city. A UNEP model for POPs management was used to assess the main anthropogenic sources of dioxin and furan. The quantity of dioxin and furan released from the identified source groups was quantified by using UNEP toolkit default emission factor. Open- and close-ended questionnaires, FGD, and key informant interview were used to collect primary information.

Secondary data were collected from various organizations such as trade and industry birroue, Addis Ababa Environmental Protection authority, Ministry of Environment, Forestry and Climate Change, Addis Ababa city municipality, and other unmentioned source category organizations. The data collected from such kind offices were analyzed and presented in the form of supplementary information for the analysis. Relevant peer reviewed research articles, international and national legislations, and related literature were collected online and reviewed systematically at informative manner. The articles and standard literatures were read critically to synthesize the information.

3. Environmental and health impacts of dioxine and furan

PCDD/F and PCB are considered dioxin- and furan-like compounds under Stockholm Convention. PCDDs and PCDFs are unintentional by-products of incomplete combustion process of chlorinated products and known to be widespread and persistent in the environment [3]. 2,3,7,8-tetrachlorodibenzo-p-dioxin (TCDD) has been known to be the most toxic congener of all PCDDs and was classified as a carcinogenic 24 substance by WHO's International Agency for Research on Cancer in 1997 [4]. Infants are more sensitive to the exposure of dioxins and dioxin-like compounds such as PCBs. Studies indicated that dioxins and dioxin-like compounds may interfere with thyroid hormone levels, increase risk of growth retardation, delay in developmental landmarks, cause neurocognitive deficits, and lead to reproductive impairments [5–8]. With short-term and high exposure level, TCDD can cause chloracne, a severe skin disease with acne-like lesions on the face and neck, which may also extend to the upper

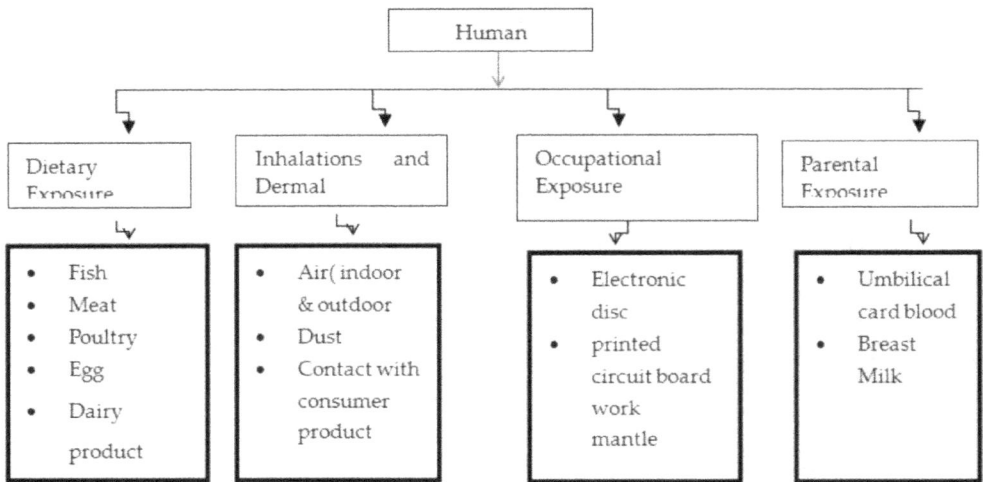

Figure 1. Overview of the human exposure pathways to POPs.

body in human [4]. With acute (14 days or less) oral exposure or intermediate (15–364 days) and chronic (365 days or more) exposure to PCDDs, liver's metabolism, biochemical, and weight change [4]. In Taiwan, several studies were done to determine the effect on children of Yu-Cheng women, who were transplacentally exposed to the pollutants. The results showed that prenatal exposure to PCBs and PCDFs could potentially result in reduced neurocognitive development, hyperpigmented skin and nails, and increased the risk of growth retardation [8]. Furthermore, Guo et al. [5] found that prenatal exposure possibly causes deterioration in semen quality, increases the percentage of abnormal sperm, and reduces daily sperm production in sexually matured men who experience in utero exposure.

In recent years, increasing concern has been given to polybrominated congeners of dibenzo-p-dioxins/furan. Brominated dioxins exert their toxic effects through the same mechanism as their chlorinated congeners, of which TCDD and its brominated congener 2,3,7,8-tetrabromodibenzo-p-dioxin (TBDD) presented almost identical potencies for the immune toxic effects on mice [9]. According to the Environmental Health Criteria 205, PBDD/Fs exhibits similar health effects to the animal as their chlorinated congener PCDD/Fs. With the oral exposure of TBDD, spermatogenic activity decreases and thyroid hormones changes in the Westar rate while growing retardation and histopathological changes in liver and thymus detects in the Sprague-Dawley rats. However, data on human exposure of PBDD/F are scarce. As it was shown in **Figure 1**, exposure ways of dioxine and furan are diverse. Only limited studies were done on these emerging pollutants, and most of them are concentrated on occupational exposure [4].

4. The birth of the Stockholm Convention

The issue of POPs begun to be addressed in the 1980s; Canada has brought it onto the international agenda [10–12]. In Canada, research found out and heightened the sensitivity of concerns of its northern indigenous populations. It was in the 1983, when the international organizations like UNEP and WHO asked Canada to turn down Canada effort to seek international actions to reduce these chemical actions. During this time, the government of Canada has entered into force the convention of management of long-range transboundary air pollution [12]. Although its implementation was regional rather than global, the fact that the convention covered most of the northern hemisphere made it a feasible forum. Ultimately, the convention integrated POPs into its agenda, it created POPs agreement that covered many of the initial "dirty dozen" of these chemicals in the country.

Then, the issue once addressed at an international scale in the form of soft law by the 1992 Rio UN Conference on Environment and Development and Agenda 21. By 1995, the UNEP Governing Council initiated an assessment process regarding a list of 12 POPs subject to future regulation [13]. An Ad Hoc Working Group on POPs was established to develop a plan for assessing information on the chemistry, environmental dispersion, toxicity, sources, and socioeconomic impacts of a list of 12 chemicals called the "dirty dozen" [13]. By 1996, the working group concluded the need for global action to include the establishment of a global, legally binding instrument. Consecutive meetings that were held from 1998 to 2001 have succeeded to produce signatory document for countries. In 2001, in the conference that was held in Stockholm, Sweden, about 91 countries and the EU signed the Convention.

5. Identification and quantification of release of dioxin and furan unintentional persistence organic pollutant

One of the major goals of the Stockholm Convention on Persistent Organic Pollutants (POPs) is the continuing minimization and, where feasible, ultimate elimination of unintentionally produced POPs. Parties are required to identify, characterize, quantify, and prioritize sources of releases of unintentionally produced POPs and develop strategies with concrete measures, timelines, and goals to minimize or eliminate these releases. Toward this end, parties must develop action plans as part of their National Implementation Plans (NIP) to identify, characterize, and address the releases of unintentional POPs Listed in Annex C of Stockholm Convention. Action plans to be developed according to Article 5 of the Convention shall include evaluations of current and projected releases that are derived through the development and maintenance of source inventories and release estimates, taking into consideration the source categories listed in Annex C of the convention tool.

According to UNEP [14], toolkit has been assembled for the purpose of assisting each country in identifying and quantifying sources of unintentional POPs that are located within the country's borders and estimating releases from those sources. Sources of POPs releases are of four general types, three of which are active, ongoing processes and one is a legacy of historic activities:

- Chemical production processes, e.g., facilities or production units that produce chlorinated phenols or in which certain other chlorinated chemicals are manufactured or that produce pulp and paper using elemental chlorine for chemical bleaching;

- Thermal and combustion processes, e.g., waste incineration, combustion of solid, and liquid fuels, or production of metals in thermal processes;

- Biogenic processes in which PCDD/PCDF may be formed from precursors-manufactured chemicals such as pentachlorophenol that is structurally closely related precursors of PCDD/PCDF.

- Reservoir sources such as historic dumps containing PCDD/PCDF and other POPs contaminated wastes and soils and sediments in which POPs have accumulated over time.

Some additional source categories and a strategy for identifying new source categories are presented in the toolkit (Annex C). It describes a step-by-step process to estimate PCDD/PCDF releases from each source category to the following environmental media:

- Air,

- Water (surface and ground water, including marine and estuarine water),

- Land (surface soils) and to these process outputs: products (such as chemical formulations, including pesticides or consumer goods such as paper, textiles, etc.) and residues (including certain liquid wastes, sludge, and solid residues, which are handled and disposed of as waste or may be recycled).

6. World experiences in POPs monitoring and management

In recent years, the potential risks of POPs were considered seriously throughout the world. As such, regulatory control and measures were proposed to protect human health, the ecosystem, and the environment. According to report of Porta et al. [15], few countries including USA, Germany, and Arctic had established nationwide surveillance programs of human concentrations of POPs, while other countries have conducted population-wide studies on the distribution and concentrations of POPs.

6.1. United States

The United States has taken several actions to control and reduce the emission of POPs. Environmental Protection Agency (US EPA) controls and manages the release of dioxins and furans to air, water, and soil by the Clean Air Act and the Clean Water Act. According to the 1990 Clean Air Act, the US EPA identifies the major industrial sources of toxic air emission and sets regulations using a technology-based or performance-based approach to reduce toxic emissions. Industries must be scaled up to achieve the maximum control of hazardous air pollutants, including dioxins and furans [16]. In addition, the act demanded EPA to analyze the remaining risks and decide whether control measures have to be tightened. Apart dioxins management using clean air act, the Clean Water Act controls and manages the release of dioxins to water through a combination of risk-based and technology-based tools [17]. Moreover, EPA Superfund and Resource Conservation and Recovery Act Corrective Action programs were also helping to clean up the dioxin-contaminated land [18]. Several conventions and commissions were established on a regional basis to control and address regional environmental concerns.

Various activities have been performed by different sectoral offices to reduce the pollution and its effect on human health. The CDC [19] indicated that National Report on Human Exposure to

Environmental Chemicals (NRHEEC) in the United States has given an opportunity for ongoing assessment of the POPs exposure level of the U.S. population. This assessment program has linked POPs to the National Health Nutrition Survey (NHANES) and determined the concentration of some POPs in the blood and urine of the general population from 50 states. The program assessed POPs increased from 27 in 2001 to 148 by 2005. POPs that were assessed in the program included the pesticide POPs (aldrin, chlordane, DDD, DDE, DDT, dieldrin, endrin, heptachlor, mirex, toxaphene, etc), brominated compounds (PBDEs, polybrominated biphenyls (PBBs), etc), and other additional chemicals in 2005. Furthermore, fluorinated POPs, such as PFOA, PFOS, and its salts, were also found under the candidate chemicals in priority groups [19]. Ecotoxicology database (ECOTOX) program was also setup by the United States Environmental Protection Agency (US EPA) to study the human exposure. The aim of the program was to consolidate all toxicity data for aquatic and terrestrial organisms from peer-reviewed literature [20]. Currently, it is funded and being managed by the U.S. Department of Defense's Strategic Environmental Research and Development Programme (SERDP) and the U.S. EPA's Office of Research and Development (ORD) and the National Health and Environmental Effects Research Laboratory's (NHEERL's) Mid-Continent Ecology Division [20]. The US EPA finally made the database to be publicly available by the Toxics Release Inventory (TRI) program to track the emission levels and trends for toxic chemicals, including dioxins and furans, from major industries, point sources, stationary sources, and mobile sources [16].

6.2. Germany

In Germany, Environmental Survey (GerES) of POPs officially started in 1985. Serious four GerESs were conducted in the period of 1985–1986 (GerES I), 1990–1992 (GerES II), 1998 (GerES III), and 2003–2006 (GerES IV), respectively [21]. Blood and urine from the general population of ages between 25 and 69 were analyzed. In GerES II and GerES IV reports, addition subsamples were done for children aged between 6 and 14 years and another aged between 3 and 14 years [22].

6.3. Arctic

In Arctic, the human monitoring on POPs was done by the Arctic Monitoring and Assessment Programme (AMAP) in 1994 and 2002 to assess the contaminant levels in the blood (AMAP, 2000; AMAP, 2002). Blood samples from human living in the Arctic regions of the eight circumpolar countries, including Canada, Denmark/Greenland, Finland, Iceland, Norway, Russia, Sweden, and the United States, were collected and analyzed for POPs in order to evaluate the regional and spatial trends of pollution, to determine the sources and its pathways, and to determine human impact, including chronic effect, posed by POPs in this region. They have drawn spatial trends by evaluating the POPs level detected in the blood samples of human from different regions.

6.4. Canada

In Canada, pesticides-related POPs are managed under the Pest Control product Act (PCPA). For management of these pollutants, all import, used, and sold activities of pesticides must be registered at the Pest Management Regulatory Agency [23]. Under the Protocols on Persistent

Organic Pollutants, the production and use of aldrin, chlordane, chlordecone, dieldrin, endrin, hexabromobiphenyl, mirex, and toxaphene are banned. Also, the use of DDT, HCH (including lindane), and PCBs is restricted. The Toxic Substances Management Policy (TSMP) of the country is a preventive and precautionary approach to virtually eliminate all POPs listed under the Stockholm Convention and to minimize the release of such chemicals to the environment [24]. Furthermore, Canada has adopted a new program called Chemicals Management Plan to evaluate the negatively impacting chemicals by risk assessment approaches for determining whether a chemical usage restriction or ban [25]. Also, the National Pollutant Release Inventory (NPRI) of Environment Canada, in its Environmental Protection Act, 1999, necessitates all company or facility which has approximately 10 or more full-time employees and uses one or more of the listed hazardous substances to register and report the total amount of each of the hazardous substances used during the year [25].

6.5. The European Union

In the European Union (EU), REACH, a new legislation which deals with Registration, Evaluation, Authorization, and Restriction of Chemical substances, entered into force on June 1, 2007 [26]. Currently, around 30,000 marketed chemicals, new or existing, required to register under REACH, in which 1500 of the chemicals are subjected to permission before their use and their introducing to the market. Chemicals which require the authorizations are those which are classified as carcinogenic, mutagenic, or toxic to reproduction compounds; persistent, bioaccumulative, and toxic compounds. The authorization will only valid for a limited period of time. Under REACH, industries are responsible for assessing and to manage the risks posed by their chemical substances [26]. Furthermore, for chemicals that are imported or manufactured more than 1 ton per year per company, manufacturers and importers are responsible for gathering information on the properties of their chemical substances to ensure safe handling of such chemicals and to register such information in a central database run by European Chemicals Agency (ECHA). Chemicals that are imported or manufactured more than 10 tons per year per company have no chemical safety report which contains the chemical safety assessment, information on the persistent nature, bioaccumulative, and toxicity behavior of the chemicals and the human health and environmental risk assessment [27]. In addition to the chemical database, the European Community and 23 Member States will implement the publicly accessible database, European Pollutant Release, and Transfer Register (E-PRTR), succeeding the European Pollutant Emission Register (EPER). E-PRTR covers the releases of pollutants to air, water, land, and off-site transfers of waste releases from diffuse sources such as road traffic and domestic heating. Facilities which emission exceeded the limited set under the regulation are obligated to report to their member states. This report will be published annually [26].

7. Management practice of persistence organic pollutant in Ethiopia

According to NIP (2006), the management practice of persistence organic pollutant is weak after the convention ratified by the country. Even though there are some activities regarding the management of this persistence organic pollutant, there are several legislations which are applicable to POPs in one way or another. Environmental Pollution Control Proclamation No. 300/2002 and Pesticide Registration and Control Council of State Special Decree No. 20/1990 are among the most important legislations for regulating POP chemicals in Ethiopia. However, analysis of the relevant legislations and their enforcement indicates that the legal system that relates to the management and use of chemicals in general and POPs in particular in Ethiopia is far from well developed. Lack of comprehensive approach and coverage is one of the major shortcomings of the legal framework. The other major gap and limitation in the area are a lack of legislations and standards in the following areas:

- Lack of rules that expressly ban the production, import and use of POPs pesticides

- Lack of legislation that directly and comprehensively regulates industrial chemicals, including PCBs;

- Lack of proper regulatory mechanism for the use of DDT;

- Lack of enabling legislations and standards to regulate releases of unintentionally produced POPs from different source categories;

- Lack of proper regulatory mechanism for the management of POPs stockpiles and wastes;

- Lack of proper regulatory framework on information gathering and exchange;

- Lack of regulatory framework on public awareness and participation

8. Main anthropogenic based dioxin and furan source categories

Contrast to natural phenomena, anthropogenic source is potential or potent for emission of persistence organic pollutant such as polychlorinated dibenzodioxin and polychlorinated dibenzofuran (PCDD and PCDF). Unintentional persistence organic pollutants (PCDD and PCDF) are produced from four general types of anthropogenic activities found in a given country and city ([2]).

- Chemical production process, e.g., facilities or production units that produce chlorinated phenol

- Thermal and combustion process, e.g., waste incineration, combustion of solid and liquid fuel, or production of metals in thermal processes:

- Biogenic process in which PCDD/PCDF may be formed from precursors-manufactured chemicals such as pentachlorophenol that are structurally closely related precursor of PCDD/PCDF

- Reservoir source such as historic dumps containing PCDD/PCDF and other pops contaminated waste and soil and sediments in which POPs have accumulated over time

However, it is impossible to found out these source categories in all countries, as well as cities at the same level. It has variations on the economic activities of the countries or cities. The detail source groups are discussed below.

8.1. Source group 1—waste incineration

Waste incineration is the predominant anthropogenic activity that contributes to the emissions of unintentional persistence organic pollutants such as dioxin and furan. Under waste incineration, five sub-categories are identified and listed such as municipal solid waste incineration, hazardous waste incineration, medical waste incineration, light fraction shredder incineration, sewage sled incineration, and animal carcass incineration [2].

8.2. Source group 2—ferrous and nonferrous metal production

Under this source groups eleven source categories were identified and listed. These source categories are iron ore sintering, coke production, iron and steel metal production, foundries, hot-dip galvanizing plants, copper production, aluminum production, lead production, zinc production, brass and bronze production, and magnesium production. Nonferrous metal production (e.g., Ni) like shredders and thermal wire reclamation and e-waste recycling are commonly found in Addis Ababa [2].

8.3. Source group 3—heat and power generation

There are five categories that are identified under this source group of heat and power generation. These are fossil fuel power plant, Biomass power plant, landfill biogas combustion, Household heating, and Cooking–Biomass and Domestic heating–Fossil fuel [2].

8.4. Source group 4—production of mineral products

Under this production of mineral product, there are six source categories activities responsible for the emission of dioxin and furan. These are cement kiln, lime production, ceramic production, brick production, glass production, asphalt mixing, and oil shale processing [2].

8.5. Source group 5—transport

Transport is one of source group that releases dioxin and furan organic pollutant due to the incomplete combustion of an engine. Under this source group, 4 stroke engine, 2 stroke engine, diesel engine, and heavy oil engine vehicles categories are responsible for emission of unintentional persistence organic pollutant [2].

8.6. Source group 6—open burning process

Due to the incomplete combustion of biomass, emission of PCDD/PCDF toward the environment Medias such as air, water, land and products will be occurred. Thus, under the group of open burning process only two categories were identified such as biomass burning (forest fire sugar cane etc.) and waste burning and accidental fires(accidental fire on vehicle and house) [2].

8.7. Source group 7—production and use of chemicals and consumer goods

According to UNEP toolkit [2], production and use of chemical and consumer goods are one of the anthropogenic activities for emission of persistence organic pollutants which have been classified under the source group of unintentional persistence organic pollutants. This source group also classified into various categories such as pulp and paper mills, chlorinated inorganic chemicals, chlorinated aliphatic chemicals, chlorinated aromatic chemicals (per ton product), other chlorinated and nonchlorinated chemical (per ton product) petroleum refining, textile plants (per ton textile), and leather plants.

8.8. Source group 8—miscellaneous

This source group has classified into five categories. The source categories listed on toolkit guide are: drying of biomass, crematoria, smokehouses, dry cleaning, and tobacco smoking.

8.9. Source group 9—disposal

Waste disposal is an anthropogenic activity which contributes to the emission of the unintentional persistence organic pollutant so-called polychlorinated dibenzodioxins and polychlorinated dibenzofuran (PCCD/PCDF). Under this source group, landfills, waste dumps and landfill mining, sewage/sewage treatment, open water dumping composting, and waste oil disposal are source categories which have major contribution for emission of dioxin and furan [2].

9. Quantifications of dioxin and furan emission from different source group and categories

Inventory of PCDD/PCDF release has been conducted after identification of all source group and source categories. According to UNEP toolkit [2], the annual release of Dioxin and furan can be determined based on the computation of emission factor and activity data.

9.1. Emission factor for PCDD/PCDF

The default emission factors presented in the toolkit are driven from a variety of data sources from laboratory experiments, peer-reviewed, and literature dedicated experiment project to governmental or institutional report. The emission factors for each class are the best estimate based. Data on technology, process characteristics, and operating practices were taken from well-documented sources of sector offices. An expert judgment was also used.

9.2. Activities rate for PCDD/PCDF

Activity rate data are very significant data for quantification of the annual release of PCDD/PCDF in the study area, in which activity rates are value in unit per year of product manufactured (e.g., steel) or feed processed (e.g., municipal waste hazardous, coal, diesel fuel etc.), annual quantities of material released (e.g., M^3 of free gas, liter of kilogram or ton of sled generated etc.) [2].

10. Management practices of dioxin and furan

Based on the Stockholm Convention on persistence organic pollutants, countries who signed the treaty toward the management of persistent organic pollutant have to design management system like BAT and BEP (best available technology and best environmental practice) to reduce and eliminate dioxin and furan organic pollutants. This agreement is not easily applicable in a developing country like Ethiopia, where the economy is still in progress and it has a major related impact during the developmental activities.

11. Case study: Ethiopia

11.1. Emissions and management practices

11.1.1. Medical waste incineration

About 50% the health care institutions in Addis Ababa have no furnace during medical waste incineration. Moreover, 100% of the currently using incinerator does not have an air pollution control system, and their medical waste incineration activity was not environmental sound and the organizations do not have disposal site to dispose the bottom ash. Therefore, the overall management practice of PCDD/PCDF from medical waste incineration is very weak.

11.1.2. Iron and steel production

The respondent of iron and steel factories was used the furnace for heating and melting purpose of the dirty and cleaned raw materials. During the production activities, all organizations do not have air pollution control system. The management practices of PCDD/PCDF emissions from iron and steel industries are very weak. The field visit to some factories confirmed that they produce their material by melting into furnaces without installing the air pollution control system and temperature control system.

11.1.3. Heat and power generation

The Dioxin and Furan is also easily emitted from household heating and cooking with biomass. The city residents commonly uses the three pit stoves, charcoal stove, and others stoves that do not have combustion control system. This emits the PCDD/PCDF from its open burning process.

11.1.4. Glass production

Glass factories that have a furnace to produce their products do not have dust control abatement and PCDD/PCDF is easily emitted to the environmental media.

11.1.5. Transport

Vehicles in the city have no emission reduction catalyst. The management of the unintentional persistent organic pollutant emitted from the vehicle through the incomplete combustion does not be managed properly.

11.1.6. Open burning process

The household heating and cooking with biomass is also one of the major contributors of dioxine and furan.

11.1.7. Textile production

Studies indicated that about 75% of the textile wastes from the industries were simply discharged to the sewerage line rather than treating and releasing into the water bodies, whereas 25% of textile wastes react to emission control [28]. This implies that majority of the industries have improper management of waste water released from their industries.

11.1.8. Leather refining

Regarding the emissions reduction management of PCDD/PCDF from leather refining industries, 7 or 99% of leather refining industry respondents revealed that their organizations have treated the wastewater via treatment plant. The rest 1% of industries do not have treatment plant but they are simply discharge into sewerage line, so that the PCDD/PCDF reduction was properly managed at leather refining industry.

11.1.9. Dry cleaning

The management practices of PCDD/PCDF from source category of dry cleaning have no optimum management practices for emission control. About 86.7% of respondent of dry cleaning industries revealed that their organization utilized chemical for dry cleaning and disposed the by-product through sewerage line, so that they easily generated and emit PCDD/PCDF without control.

11.1.10. Tobacco smoking

There are no management practices of PCDD/PCDF emission from tobacco smoking activities in the city.

11.1.11. Waste disposal

In the city of Addis Ababa, there is no management practices of leachate or seepage liquids disposed from solid waste. This implies that PCDD/PCDF can be easily emitted from these anthropogenic activities. Also, there is no adequate treatment of sewage that can control emission of dioxin and furan from sewage treatment.

11.2. Institutional policy and regulatory framework

Regarding the policies and legal framework for persistent organic pollutant, the researcher conducted an interview with higher expert working in Ministry of the environment, forest and climate in compliance and Monitoring directorate, and then some question were raised about the policy and regulatory framework of persistence organic pollutant. Expert from the FDRE Ministry of Environment, Forestry and Climate Change has explained that Ethiopian has designed various policy and regulatory frameworks to govern the environment, e.g., environmental policy approved by the minister council, and this policy has regulatory framework these are, environmental Protection Organs establishment, environmental pollution control proclamation (300/2002) environmental Impact Assessment Proclamation (299/2002) and etc. However, all regulatory frameworks directly or indirectly influence or contribute for the management of persistence organic pollutants, but it is impossible to say that this is good enough for emission reduction of POPs. Therefore, these can be put as the gap and weakness on the existing legislative system for the management of POP chemical in Ethiopia, as well as in Addis Ababa city. In addition, the respondent described that at present there is no legislation that specifically deals with POPs chemicals other than pesticides. The only particular legislations are those which control and regulate pesticides. Therefore from the point views of the interview, it is implied that the management practices of the emission reduction of the persistence organic pollutants through legislative and regulatory framework have not been formulated.

12. Conclusion

The major anthropogenic sources of emission of dioxin and furan (PCDD/PCDF) have generalized as 9 source groups, 15 source categories, and 25 classes. To quantify the release of PCDD/PCDF, an activity rate data and emission factor of each source category is required. The default model of emission factor prepared by UNEP to convert the annual activity rate data into annual release OF PCDD/PCDF in g TQE for different environment media is mostly used if there is no experimentally proved emission factor data found.

Most of the service sectors in developing countries have no control system for emission of dioxin and furan. Also, the regulatory issue of dioxin and furan is risen in various stages of the environment policy and strategies; however, there is no separate guidelines and standards prepared for it. The management practice of dioxin and furan in developing countries is not satisfactory, so that it needs consideration technology, capacity building, and regulatory empowerment options.

12.1. Technology options

- Constructing a standard medical waste incinerator station with all facilities at a specific station from collecting, hauling, and transporting the medical waste from all the health care institution to the incinerator station. These flows of management highly reduce the emission of PCDD/PCDF from the uncontrolled medical waste incineration which undertaking at different health care institution in the city.

- Adoption of best available technology and best environmental practices in leather, textile, and minerals production processes to reduce or eliminate releases of PCDD/PCDF through detailed assessment of individual industries for BEP options for UPOPs reduction and need and introduce and effectively implement guidelines on BAT and BEP to release sources of UPOPs (existing and new industry)

• Removal of barrier of introduction of technologies that minimize UPOPs through environmentally sound management practices

12.2. Capacity building option

• Conduct awareness raising and establishing network for information exchange through sensitizing the public and stakeholders on environmental and health impact of dioxin and furan

• Develop education and awareness materials on health and environmental effects of UPOPs,

• Establish free access Web and database on dioxin and furan.

12.3. Regulatory framework option

• To control and reduce the release of dioxin and furan through various environmental media, the countries need to work collaboratively in stakeholders with the regulatory formulating bodies.

Conflict of interest

There is no conflict of interest with all the content of this chapter.

Author details

Mekonnen Maschal Tarekegn[1]* and Efrem Sisay Akele[2]

*Address all correspondence to: maschalm12@gmail.com

1 Department of Environment and Climate Change Management, Ethiopian Civil Service University, Addis Ababa, Ethiopia

2 Addis Ababa Solid Waste Recycling and Disposal Project Office, Addis Ababa, Ethiopia

References

[1] Ritter L, Solomon KR, Forget J. Persistent Organic Pollutants; 1995. Available from: http://www.chem.unep.ch/pops/ritter/en/ritteren.pdf

[2]   UNEP. Stockholm Convention; 2013. Available from: http://chm.pops.int/TheConvention/ThePOPs/tabid/673/Default.aspx

[3]   SC POPs. Stockholm Convention on Persistent Organic Pollutants; 2008. Annex D

[4]   WHO, World Health Organization. Polybrominated dibenzo-p-dioxins and dibenzofurans. Environmental Health Criteria. 1998;**205**. Available from: http://www.inchem.org/documents/ehc/ehc/ehc205.htm

[5]   Guo YL, Hsu PC, Hsu CC, Lambert GH. Semen quality after prenatal exposure to polychlorinated biphenyls and dibenzofurans. The Lancet. 2000;**356**:1240-1241. DOI: 10.1016/S0140-6736(00)02792-6

[6]   Jacobson JL, Jacobson SW. Intellectual impairment in children exposed to polychlorinated biphenyls in utero. New England Journal of Medicine. 1996;**335**:783-789. DOI: 10.1056/NEJM199609123351104

[7]   Pluim HJ, De Vijlder JJM, Olie K, Kok JH, Vulsma T, Van Tijn DA, Van Der Slikke JW, Koppe JG. Effects of pre- and postnatal exposure to chlorinated dioxins and furans on human neonatal thyroid concentrations. Environmental Health Perspectives. 1993;**101**:504-508. Available from: https://www.ncbi.nlm.nih.gov/pubmed/8137779

[8]   Rogan WJ, Gladen BC, Hung KL, Koong SL, Shih LY, Taylor JS, Wu YC, Yang D, Ragan NB, Hsu CC. Congenital poisoning by polychlorinated biphenyls and their contaminants in Taiwan. Science. 1988;**241**:334-336. Available from: https://www.ncbi.nlm.nih.gov/pubmed/3133768

[9]   Ao K, Suzuki T, Murai H, Matsumoto M, Nagai H, Miyamoto Y, Tohyama C, Nohara K. Comparison of immunotoxicity among tetrachloro-, pentachloro, tetrabromo- and pentabromo dibenzo-p-dioxins in mice. Toxicology. 2009;**256**:25-31. DOI: 10.1016/j.tox.2008.10.024

[10] Fenge T, POPs and Inuit: Influencing the Global Agenda. Northern Lights Against POPs. In: Downie DL, Fenge T, editors. Combating Threats in the Arctic. Montreal, Canada: McGill-Queen's University Press; 2003

[11] Kohler P, Ashton M. Paying for POPs: Negotiating the Implementation of the Stockholm Convention in Developing Countries. International Negotiation. 2010;**15**:459-484. DOI: 10.1163/157180610X529636

[12] Selin H, Eckley N. Science, politics, and persistent organic pollutants. The role of scientific assessments in International Environmental Co-operation. International Environmental Agreements: Politics, Law and Economics. 2003;**3**:17-42

[13] ENB. Summary of the First Conference of the Parties to The Stockholm Convention. Earth Negotiations Bulletin; 2005. Vol. 15(117). Available from: http://enb.iisd.org/vol15/enb15117e.html

[14] UNEP. Stockholm convention [Online]; 2013. Available from: http://chm.pops.int/TheConvention/ThePOPs/tabid/673/Default.aspx

[15] Porta M, Puigdomenech E, Ballester F, Selva J, Ribas-Fito N, Llop S, Lopez T. Monitoring concentrations of persistent organic pollutants in the general population: The international experience. Environment International. 2008;**34**:546-561

[16] US EPA. U.S. Environmental Protection Agency. Toxic Air Pollutants; 2007. Available from: http://www.epa.gov/air/airtrends/aqtrnd95/tap.html

[17] US EPA. U.S. Environmental Protection Agency: Persistent Organic Pollutants: A Global Issue, A Global Response; 2008. Available from: https://www.epa.gov/international-cooperation/persistent-organic-pollutants-global-issue-global-response

[18] US EPA. U.S. Environmental Protection Agency. Superfund: Montana Pole and Treating Plant; 2009. Available from: http://www.epa.gov/region8/superfund/mt/montana_pole

[19] CDC, Department of Health and Human Services-Centers for Disease Control and Prevention; 2003. Chemical nomination process. Available from: http://www.cdc.gov/exposurereport/chemical_nominations.htm

[20] US EPA. U.S. Environmental Protection Agency: Pesticide Environmental Stewardship Program; 2009. Available from: http://www.epa.gov/oppbppd1/pesp/

[21] Umweltbundesamt. Health and Environmental Hygiene. In: German Environmental Survey (GerES); 2009. Available from http://www.umweltbundesamt.de/gesundheit-e/survey/index.htm

[22] Umweltbundesamt. Health and Environmental Hygiene–German Environmental Survey for Children 2003/6–GerES IV–Human Biomonitoring–Levels of selected substances in blood and urine of children in Germany; 2008. Available from: http://www.umweltbundesamt.de/gesundheite/survey/kinder.htm

[23] Environment Canada. Chapter 2. Canada and POPs; 2005. Available from: http://www.ec.gc.ca/ceparegistry/documents/nip/nip_part1/Part1_p2.cfm

[24] Environment Canada. Chlorinated Substances Action Plan—Executive Summary; 2002. Available from: http://www.ec.gc.ca/nopp/docs/rpt/csap/en/summary.cfm

[25] Environment Canada. National Pollutant Release Inventory—Tracking Pollution in Canada. 2009. Available from: http://www.ec.gc.ca/pdb/npri/npri_home_e.cfm

[26] European Commission. Environment Chemicals—REACH. 2009. Available from: http://ec.europa.eu/environment/chemicals/reach/reach_intro.htm

[27] Kemmlein S, Herzke D, Law RJ. Brominated flame retardants in the European chemicals policy of REACH-Regulation and determination in materials. Journal of Chromatography A. 2009;**1216**:320-333. DOI: 10.1016/j.chroma.2008.05.085

[28] Akele ES, Tarekegn MM. Assessment of dioxin and furan emission levels and management practices in Addis Ababa, Ethiopia. 2017;**8**(15):85-94. DOI: 10.5696/2156-9614-7.15.85

Mechanistic Considerations on the Hydrodechlorination Process of Polychloroarenes

Yoshiharu Mitoma, Yumi Katayama and
Cristian Simion

Additional information is available at the end of the chapter

http://dx.doi.org/10.5772/intechopen.79083

Abstract

Defunctionalization of organochlorines through reductive dechlorination (also known as hydrodechlorination—replacement of chlorine atoms by hydrogen—is one of the main methodologies used in the detoxification of these harmful compounds. Most of the published papers on this particular matter focused on specific reagents, reaction conditions, and mainly result efficiency. Some of the authors were also concerned with reaction pathways (e.g., the order in which chlorine atoms were removed from a polychlorinated aromatic substrate—polychlorinated biphenyls, PCBs; polychlorinated dibenzo-p-dioxins, PCDDs; or polychlorinated dibenzofurans, PCDFs). However, the papers that dealt with the investigation of reaction mechanism were rather scarce. This chapter presents the advances made by researchers in understanding, from a mechanistic point of view, the hydrodechlorination process, along with our own assumptions. In doing so, it would be easier to predict the behavior of such compounds in a specific environment, showing more clearly the scope and limitations of each process, depending on the reaction conditions and reagents.

Keywords: hydrodechlorination, reaction mechanism, metal/hydrogen donor

1. Introduction

Most chlorinated and especially polychlorinated arenes (such as polychloro-dibenzo-*p*-dioxins **1**, polychloro-dibenzofurans **2**, or polychlorobiphenyls **3**) are persistent organic pollutants (POPs) that are harmful to both man and the environment [1–5].

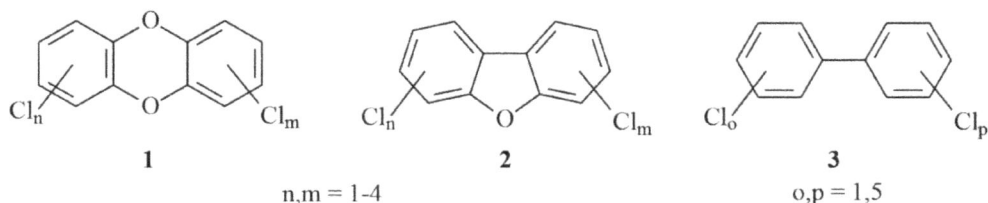

1

n,m = 1-4

2

3

o,p = 1,5

As a result, numerous techniques and procedures were implemented for their destruction/ degradation [6–11]. Initially, these procedures focused on the separation/extraction of poly-chlorinated compounds and subsequent treatment. Nowadays, researchers value the in situ procedures the most, with the advantage of time and the economic aspects [7, 12–14]. Indeed, such compounds are stable molecules, which are resistant to hydrolysis, oxidation, and tem-perature changes, thus being difficult to degrade and showing long half-life times in the envi-ronment [15]. For example, dioxins could have atmospheric half-lives of 10–20 years, while, in soils, they can reach staggering values of up to 150 years.

This means that it might be an impossible task to deliver a dioxin-free Earth for future genera-tions, but it lies on the present generation of scientists and engineers to try to do so. In that spirit, early attempts appealed to mankind's most powerful discovery of destructive technol-ogy—fire [16]. In the early 1990s, most of the processes were thermal. But as it was soon discovered, these procedures were in fact a source for a de novo polychlorinated compound synthesis, the process by which dioxins are re-formed being investigated in the recent years [17–19]. Therefore, greater attention was focused on more chemical-based processes [6, 10, 20-23], along with microbial ones [24].

In particular, attention was focused on those procedures that were based on the combina-tion of a metal and a hydrogen donor, processes that allowed the hydrodechlorination of the polychlorinated aromatic substrate to less toxic hydrocarbons [6–9, 14, 21–23]. Since these are among the most studied procedures in the past few years, we turned our attention to the particular reaction mechanism of the hydrodechlorination reaction. Indeed, understanding the reaction mechanism of a particular process is important in view of understanding the role played by each reagent but also in view of the predictive modeling of similar processes or the same process applied to various other substrates [25].

2. Proposed hydrodechlorination mechanisms–Literature survey

From an historical perspective, the formation of polychlorinated arenes (PCDDs, PCDFs, and PCBs) were first investigated [23, 26–34] and soon the catalytic role of various metals was understood [35–41]. If metals can do, they can also undo: indeed, they can effectively intervene in numerous defunctionalization processes, including hydrodechlorination [22, 42]. Since the process implies a reductive approach, researchers naturally oriented themselves toward reduction catalysts, such as Ru, Rh, Pd, or Pt. One remark must be made for the pio-neering efforts of Ukisu and coworkers [43–46], who used a hydrogen donor (isopropanol),

NaOH, and Pd/C or Ru/C. Ukisu also provided the first insights on the possible reaction mechanism of the hydrodechlorination process, apprehending the intervention of atomic (or nascent) hydrogen:

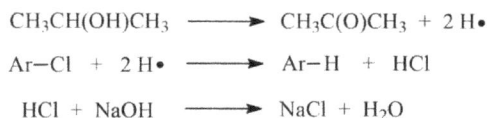

$$CH_3CH(OH)CH_3 \longrightarrow CH_3C(O)CH_3 + 2 H\bullet$$

$$Ar-Cl + 2 H\bullet \longrightarrow Ar-H + HCl$$

$$HCl + NaOH \longrightarrow NaCl + H_2O$$

Later on, Ukisu also tried to explain the intervention of the catalyst. Based on much earlier studies, Ukisu stated that the dehydrogenation of 2-propanol to acetone on rhodium complexes implied the elimination of a hydride ion [47]—Ukisu assumed that the α-hydrogen of isopropanol transfers to PCDD/F in the form of hydride, on the catalyst's surface [45, 46], the reaction resembling to an aromatic nucleophilic substitution (**Figure 1**):

The trail of reduction catalysts is still a heavily investigated one. Ayame's team started from the premise that, in the catalytic hydrodechlorination of monochlorobenzene, the hydride ion formed on the Pd surface spills over the alumina carrier surface and attacks the electron-deficient carbon of a monochlorobenzene adsorbed on Lewis acid sites of the alumina to produce a benzene molecule and a chloride ion. The chloride ion, which coordinated to a Lewis acid site, would be converted to hydrogen chloride in the reaction with H+ spilled over from the Pd surface [48]. However, most scientists were at that point more interested in the various pathways with regard to the reactivity of the differently positioned chlorine atoms [49–54]. Based on density functional theory calculations, rules of thumb for assessing the reductive dechlorination pathways of PCDDs were proposed [55]. These included "(1) the chlorine atoms in the longitudinal (1,4,6,9) positions are removed in preference to the chlorine atoms on lateral (2,3,7,8) positions; (2) the chlorine atom that has more neighboring chlorine atoms at ortho-, meta- and para-positions is to be eliminated; (3) reductive dechlorination prefers to take place on the benzene ring having more chlorine substitutions; and (4) a chlorine atom on the side of the longitudinal symmetry axis containing more chlorine atoms is preferentially eliminated" [55].

Anyway, one important aspect resulted from these studies, and that is, that adsorption phenomenon on the catalyst's surface certainly play a role in the process [49, 56, 57]. Nevertheless, there was no consensus on the hydride transfer mechanism. If such transfer seems plausible when an alcohol is used as hydrogen donor [43–46, 54], and even if the formation of hydride

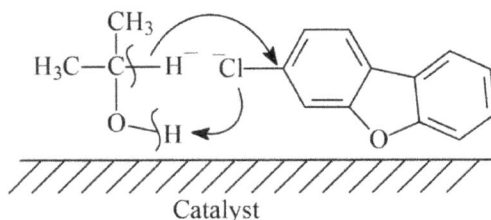

Figure 1. Ukisu's hydrodechlorination mechanism proposal [45, 46].

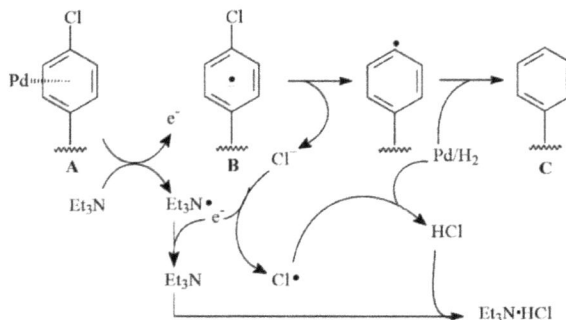

Figure 2. Tentative mechanism of the hydrodechlorination of aromatic chlorides [58].

is postulated when molecular hydrogen is involved (through a heterolytic dissociation of H_2) [48], for Hirota and coworkers, the reaction mechanism involves a single-electron transfer (SET) step [58]. This team used triethylamine (instead of NaOH) for HCl trapping, its role being also an activator of the Pd/C-catalyzed hydrodechlorination process. Upon addition to the hydrodechlorination reaction mixture of small amounts of tetracyanoethylene or 7,7,8,8-tetractanoquinodimethane, which are well-known electron scavengers, the reaction was suppressed, suggesting thus a single-electron transfer (SET) mechanism (**Figure 2**).

The SET mechanism was reprised in several other papers in which the reaction conditions were more suitable for such a process: either photochemical [59, 60] or electrochemical [61]. The radicalic mechanism was also the center of hydrodechlorination processes, involving more active metals such as Na or K [62, 63], the main argument being the recorded formation of compounds such as quarterphenyls (as results of Fittig-Wurtz-type coupling). Another testimony for the single-electron transfer mechanism came from the study of a dual depolluting process for industrial wastewaters simultaneously polluted with chlorinated compounds and nitrites/nitrates. The authors observed a competition for electrons between reductive dechlorination and denitrification—NO_3 is reduced to NH_4^+ retarding the dechlorination due to the competition for electrons [64].

Nevertheless, the use of metals in the palette of hydrodechlorination processes took a new turn with used metals such as iron, zinc, magnesium, or calcium [65, 66]. Again, for these techniques, the electron transfer from the metal, with subsequent formation of nascent hydrogen through reaction with a proton donor prevailed, the process being summarized as:

$$Metal^0 + RX + H^+ \longrightarrow Metal^+ + RH + X^-$$

Among the metals tested, iron occupies a place of choice, used either alone [66–68] or in combination with other metals, especially Pd or Pt [57, 66, 69–74]. Three mechanisms were proposed to explain the observed dechlorination process [66]: one that involves direct electron transfer from Fe to the adsorbed alkyl halide ($Fe^0 + RX + H^+ \rightarrow Fe^{2+} + RH + X^-$) and other two that involve corrosion of Fe in water under anaerobic conditions ($Fe^0 + 2H_2O^+ \rightarrow Fe^{2+} + H_2 + 2OH^-$). The idea of using a bimetallic system, Pd along with Fe, made the process much more effective (**Figure 3**):

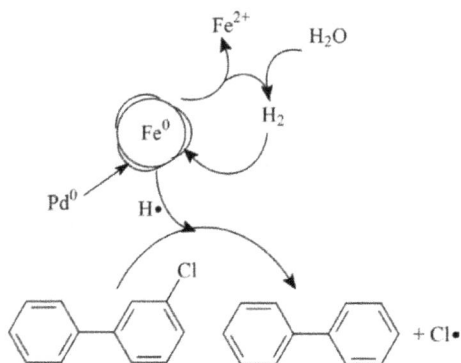

Figure 3. Proposed surface reaction of PCBs with Fe/Pd nanoparticles [74].

An interesting twist to the method is represented by the replacement of Fe with Mg in Fe/Pd [75–78], which is based on the following reasons: Mg has a relatively high oxidation potential (2.37 V; Fe has just 0.44 V), providing thus a greater thermodynamic force, and, while Fe tends to rather rapidly corrode, Mg can form a protective magnesium oxide shell (**Figure 4**).

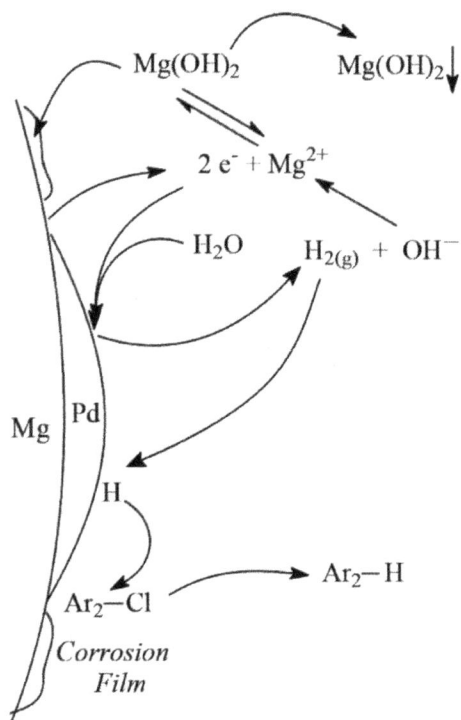

Figure 4. Proposed mechanism for the dechlorination of PCBs over Mg/Pd [76].

A comparison of the hydrodechlorination effectiveness of Pd-deposed metals (Mg, Al, Mn, Zn, Fe, Sn, and Cu) was made by Yang et al. [79], who found that, in acidic aqueous solution, Mg/Pd present the highest reactivity, while at the other end are Sn and Cu that showed little to none dechlorination capability.

An important breakthrough in understanding the reaction mechanism of palladized magnesium-mediated hydrodechlorination was made by the Geiger team [80–83]. By working in pure methanol, the source of H_2 is the reaction between Mg and the alcohol ($Mg^0 + 2CH_3$ $OH \rightarrow Mg(CH_3O)_2 + H_2$). The reaction exhibits pseudo-first-order kinetics, and the order of dechlorination rates of chlorine atoms in 100% methanol is ortho->para->meta-positions, which differ from that in water/methanol (9:1 v/v) or 100% water (para->meta->ortho-). But, more importantly, the possible reaction mechanisms were discussed. This team proposed either an $S_{RN}1$ mechanism or an $S_{RN}1$ type of mechanism, in which a nucleophilic substitution, involving a radical intermediate, occurs. One major difference from classical $S_{RN}1$ mechanism though is that the initiation step is the homolytic rupture of the C-Cl bond by either atomic hydrogen or hydride ion, with subsequent formation of aryl radical and HCl. In the more classical $S_{RN}1$ mechanism, an aryl radical anion is formed due to hydride formed on bulk Pd. Basically, the three proposed mechanisms differ by the nature of the hydrogen involved: simple radicalic H (or atomic/nascent H), atomic H with an enhanced negative charge (a radical-anion H with a fractional negative charge – $H^{\delta-}$), or a hydride ion (H^-) (**Figure 5**).

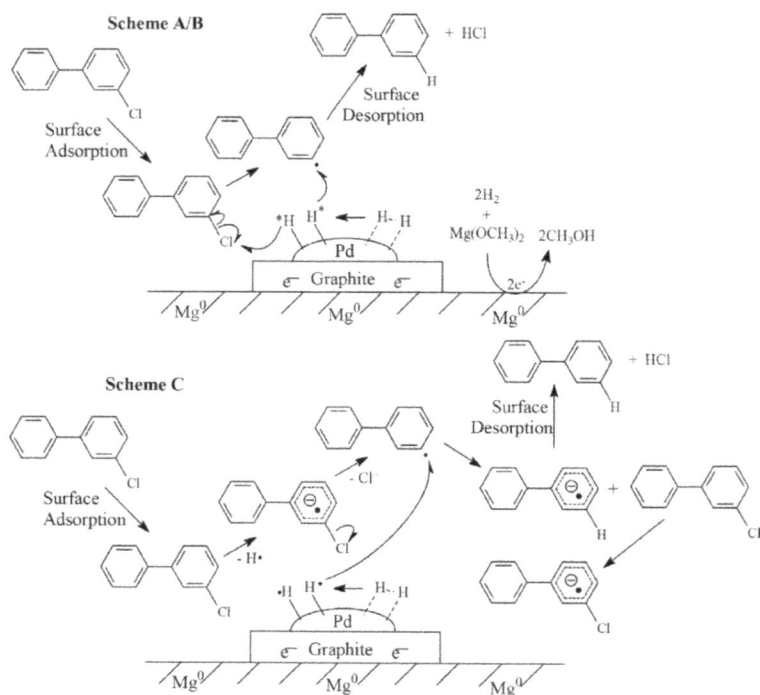

Figure 5. Proposed mechanism for the dechlorination of PCBs by Mg/Pd in methanol by scheme A/B atomic hydrogen or "hydride-like" radicals, and scheme C hydride (H^- denotes both hydrogen and "hydride-like" species) [82].

An interesting assumption for the hydride mechanism is that the latter act as a nucleophile, which can transfer an electron to the aromatic chlorinated substrate, which will cause the elimination of a chlorine atom, leaving an aryl radical. This aromatic radical can quickly react with another H⁻. The charged biphenyl species can than transfer an electron to another PCB substrate, so that process can continue to propagate.

Among the reasons given for the possibility that all three mechanisms could occur are the fact that both atomic hydrogen and hydride species can be formed on Pd from molecular hydrogen and the lack of dimerization products (quarterphenyls) or additional chlorinated by-products. These mechanisms require initial adsorption of the PCBs onto the surface of the bimetallic system, then reaction at the interface of the palladium and graphite, limiting thus considerably the mobility of the aryl radical. This limited mobility and the abundance of atomic hydrogen, hydride-like radicals, and hydrides on the surface of the catalyst almost certainly allow for the reaction of the aryl radical and second nucleophilic hydrogen rather than two separate aryl radicals coming into contact [82].

For similar reasons, Ca was used instead of Mg [84–91]: not only does Ca have an even higher thermodynamic driving force when compared to Fe or Mg ($0.44\,V-Fe$, $2.37-Mg$, $2.87-Ca$) but also $CaCO_3$ coating both protect metallic Ca surface and is easily removed, allowing this procedure to be applied both ex situ and in situ (**Figure 6**).

Two potential mechanisms were proposed over the years: one purely radicalic, that involves nascent hydrogen, and one radicalic but which implies a pseudo-nucleophilic substitution mechanism in which the addition of electrons from calcium ($Ca \rightarrow Ca^{2+} + 2e^-$) transforms the aromatic ring into a radical-anion that rapidly expels chlorine atoms. The hydrogen atom on the hydroxyl group of the alcohol is then added to the radical anion yielding a hydrodechlorinated substrate and an alkoxide (**Figure 7**).

Upon working with deuterated methanol (CH_3OD), the authors observed both the formation of deuterodechlorination products and aryl dimers, suggesting the intermediacy of the radicalic aryl species [84].

Recent results based on zeta potential determinations and hydrodechlorination reactor's internal pressure monitorization suggested that the radicalic process could be favored. Indeed, no pressure rise is a hint that molecular H_2 is not formed in the reaction $Ca^0 + 2CH_3OH \rightarrow Ca(CH_3O)_2 + H_2$

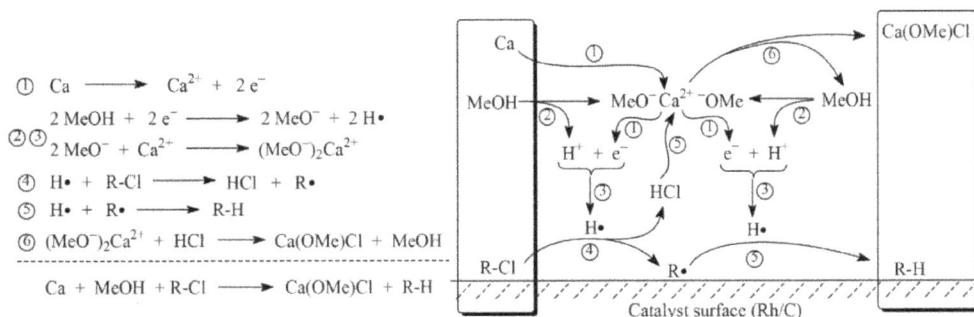

Figure 6. Possible dechlorination pathway [88].

Step 1: Ca $\longrightarrow$ Ca^{2+} + 2 e$^-$

Step 2: e$^-$ + Ar-Cl $\longrightarrow$ Ar• + Cl$^-$

e$^-$ + R-OH $\longrightarrow$ RO$^-$ + H• or 1/2 H$_2$

R = Me, H

Step 3: Ar• + H• $\longrightarrow$ Ar-H

Cl$^-$ + Ca^{2+} + HO$^-$ $\longrightarrow$ Ca(OH)Cl

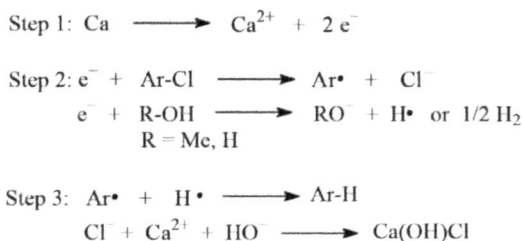

Figure 7. Hydrodechlorination mechanism following a pseudo-nucleophilic aromatic substitution [87].

but instead nascent hydrogen is generated: Ca^0 + 2CH$_3$OH → Ca(CH$_3$O)$_2$ + 2[H]. Thus, the authors assumed that during the hydrodechlorination process of chloroanisole, the only proton source is the alcoholic (protic) hydrogen, the mechanism being most probably radicalic (the recorded formation of biphenyls as coupling products is an indication for this). Even at lower H$_2$ pressures, the hydrodechlorination of chloroanisole to anisole was achieved in at least 95% or even higher yields. The reaction efficiency implies that the transfer of atomic hydrogen (formed in calcium reaction with methanol) to catalyst surface (such as Pd/C) proceeded directly, without the formation of molecular hydrogen in solution. The surface of the catalyst showed differential conditions electrostatically, depending on the concentration of calcium as electron source (**Figure 8**).

Along with metallic Ca, other Ca compounds were tested in a hydrodechlorination process: Ca(OH)$_2$ [92], CaO [93–95] or CaSiO$_3$ [95]. Although most of these studies favored a radicalic mechanism, Gao and coworkers [95] discussed alternative pathways that involved either an electron transfer (with subsequent formation of a radical-anion that expels the chlorine anion and form the aryl radical) or a direct hydrogen transfer (in an S$_{RN}$1-type mechanism). Based on calculations of the adiabatic electron affinities of PCDFs, the chloride ion dissociation yielding aryl radicals is considered the major pathway of chlorine abstraction [96]. Moreover, when traces of Cu are present, Ullman-type coupling products can be formed.

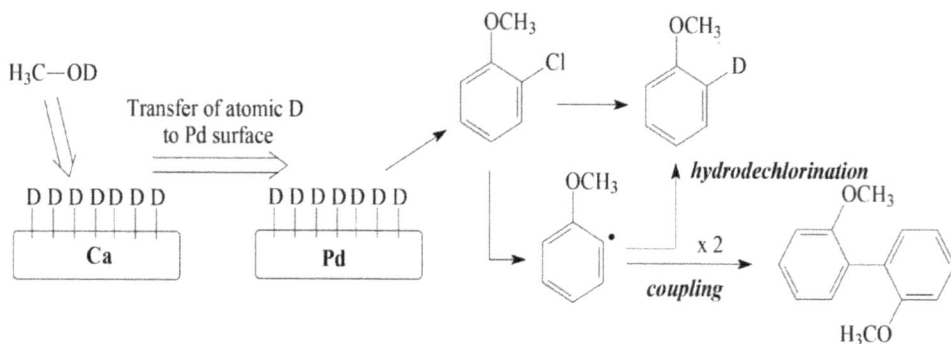

Figure 8. Proposed reaction pathway and deuterium route from H$_3$C-OD to deuterodechlorination product [84].

Similar mechanistic observations were made by different teams that used metals other than Fe, Mg, or Ca. For example, for a zero-valent Zn hydrodechlorination of tetrachlorophenol, the intermediate in the reaction mechanism is again the aryl radical formed by expelling the chlorine anion from the initial radical-anion [97]. The latter is considered to be formed by the aromatic ring quenching of an electron from zero-valent Zn $(1/2Zn^0 \rightarrow 1/2Zn^{2+} + 1\ e^-)$ (**Figure 9**).

Analogous conclusions were drawn for bimetallic systems such as Fe/Ni [98], Mg/Zn [99], or Ni/Mo [100].

Another approach was taken by Lim and coworkers when comparing the performances of Pd/Fe nanoparticles with other bimetallic systems such as Pt/Fe, Ni/Fe, Cu/Fe, and Co/Fe nanoparticles during the hydrodechlorination of trichlorophenol [72], since the process was studied in an aqueous environment. Starting from the premise that chlorinated organic compound treatment in water were reduced according to three different mechanisms—(1) direct reduction on fresh zero-valent iron (ZVI) surface, (2) reduction by ferrous iron, and (3) reduction by H_2 through catalysis [101]—the authors assumed that the bimetal/water system, having a stronger reductive ability, allowed more easily the formation of an activated reducing species, atomic hydrogen (H^*) (**Figure 10**).

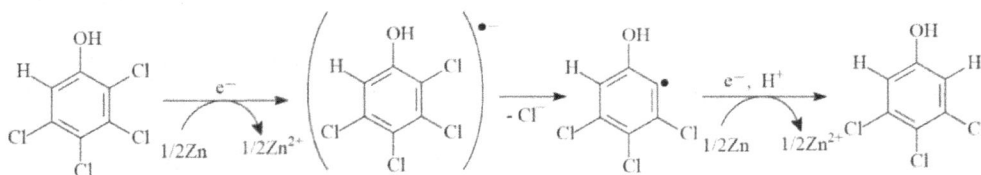

Figure 9. Hydrodechlorination of tetrachlorophenol with zero-valent Zn [97].

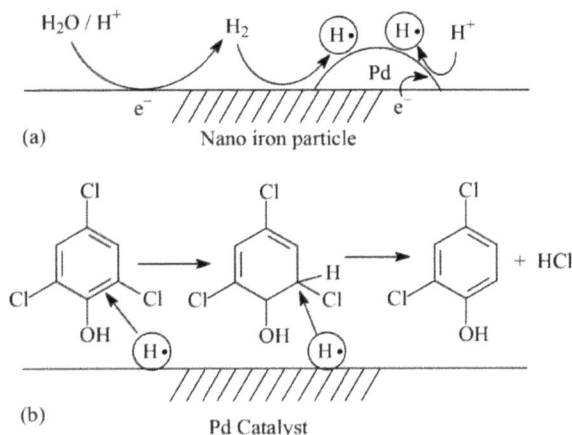

Figure 10. Schematic of proposed catalytic hydrodechlorination mechanism of chlorophenols over nanoscale Pd/Fe (a) production of atomic hydrogen and (b) surface-mediated hydrodechlorination of chlorophenols on Pd surface [72].

The production of H* may follow two routes: catalyzed decomposition of H_2 gas to H* and electron abstraction by H^+. For this particular process, the authors considered more plausible electrophilic addition, followed by subsequent elimination of HCl through a dehydrodechlorination process.

But when considering that the degradation process occurs in water or even in supercritical water, true aromatic nucleophilic substitution of the chlorine atoms (and their replacement by –OH moieties) could be considered [102, 103]. In just supercritical water and under oxidative conditions, the hydroxylated PCB only accounted for less than 10% of the reaction mass, while for alkaline, non-oxidative conditions the formation of hydroxylated PCBs could nearly close the mass balance in the early stage of PCB degradation [102]. The authors concluded that the formation of comparable amounts of PCDFs (requiring two oxygen substitution steps under preservation of both aromatic systems) in the experiments under oxidative and non-oxidative, alkaline conditions indicated that under both treatments, an oxygen substitution under preservation of the aromatic rings is the major initial step. On the other hand, hydroxylated PCBs are less susceptible to nucleophilic substitution compared to PCBs due to the electron-donating effect of the –OH group. Therefore, under the alkaline, non-oxidative conditions, the initial degradation products (hydroxylated PCBs) are less reactive toward further degradation compared to the starting compounds (PCBs). But when the treatment occurs in the presence of Co_3O_4, the reaction pathway, involving the formation of chlorophenolate and dichlorophenolate, is similar to the Mars Van Krevelen mechanism [104]. Nucleophilic substitution of the chlorine atom occurs through attack of the lattice oxygen atoms (O_2^-) and a Co-Cl bond is formed, yielding chlorophenolate and dichlorophenolate as partial oxidation products. Similar reaction routes have been reported for the formation of phenolates during the degradation of HCB over Al_2O_3 [105] and the degradation of chlorobenzene over iron and titanium oxide catalysts [106] (**Figure 11**).

An even more suggestive hint of a nucleophilic substitution of chlorine atoms was represented for the catalytic degradation of PCBs with Ni complexes [107]. A simplified model of the successive reactions was presented, clearly suggesting in steps 3 and 4 a nucleophilic attack of a hydride ion upon the aromatic chlorinated substrate:

Ivr4321

1. Ligand displacement:

$NiL_4 \rightarrow NiL_3 + L$

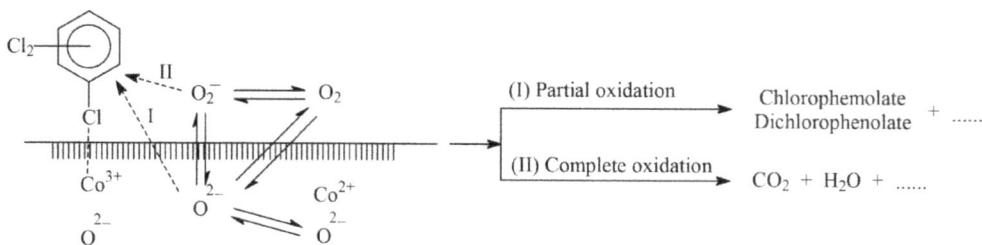

Figure 11. Oxidative attack mechanism for 1,2,4-trichlorobenzene degradation over cabbage-like Co_3O_4 [103].

2. Oxidative addition:

$$C_{12}Cl_{10} + NiL_3 \rightarrow C_{12}Cl_9NiL_2Cl + L$$

3. Reductive elimination:

$$C_{12}Cl_9NiL_2Cl + H^- + L \rightarrow C_{12}Cl_9H + Cl^- + NiL_3$$

4. Partial dechlorination:

$$C_{12}Cl_{10} + nH^- \rightarrow C_{12}Cl_{10-n}H_n + nCl^-$$

5. Complete dechlorination

$$C_{12}Cl_{10} + 10H^- \rightarrow C_{12}H_{10} + 10Cl^-$$

A study of polychlorinated biphenyls' reactivity in nucleophilic versus electrophilic substitutions demonstrated that SN reactions are of the hard acid-hard base type [108], having a lesser probability to occur.

Meanwhile, several other procedures seem to favor the radicalic hypothesis: for example, PCBs are more rapidly hydrodechlorinated when in the presence of a mixture of persulfate and quinines, a system known to generate free radicals [109]. Ultrasonication of a polychlorinated biphenyls also seem to lead to a radicalic mechanism, since the first step is presented as a homolytic cleavage of the C-Cl bond [110].

Special attention must also be given to a new entry in panoply of degradation procedures: the mechanochemical approach in the hydrodechlorination/destruction of polychlorinated arenes [11]. The mechanochemical reactions imply activation of the chemical bonds through deformation under mechanical stress, leading to their rupture and subsequent reformation [111]. Even if most mechanochemical processes applied to polychlorinated aromatic compounds implied radical intermediates, it was not uncommon to consider the formation of ionic species. For example, Birke et al. suggested one reductive dechlorination process with alkali metals in the presence of hydrogen donors—the formation in the first step of a radical anion [112] (**Figure 12**):

Generally, the activation of the substrate takes place on the solid reagent, one of the most used being CaO [113–117]. After the cleavage of CaO crystals and the exposure of new surfaces, the

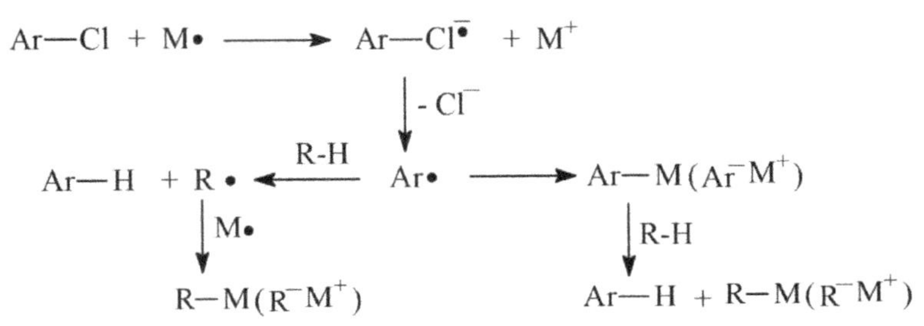

Figure 12. Reductive dechlorination of chlorinated aromatics (arylchlorides, Ar-Cl) by alkali metals in solution in the presence of hydrogen donors (R-H) like ethers or alcohols [112].

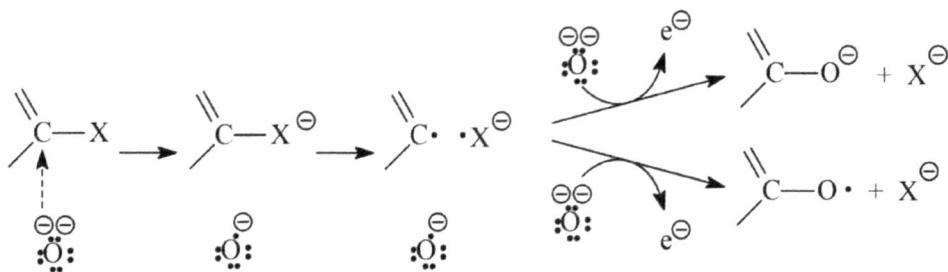

Figure 13. Dehalogenation mechanism with CaO [11].

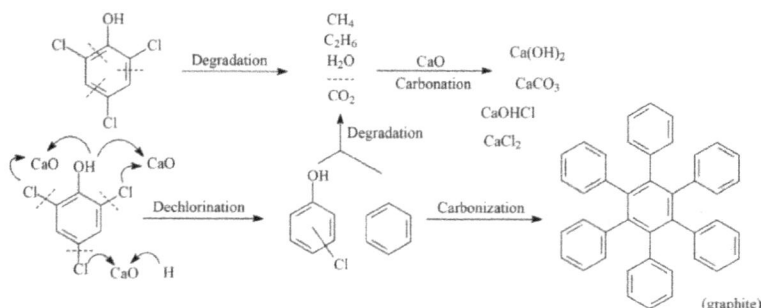

Figure 14. Degradation ways of 2,4,6-trichlorophenol in the mechanochemical reaction [117].

oxide ions are induced by mechanical activation to transfer an electron to pollutant's carbon atoms, thus generating an anion radical [11] (**Figure 13**).

The process does not even need the presence of a H donor, ultimate products being either oxidation one (e.g. CO_2) or of a graphitic nature – through carbonization processes [117] (**Figure 14**).

The addition to the reaction mixture of a good radical generator, such as SiO_2, accelerates the carbonization process [115, 118, 119]. The presence of zero-valent metals in the ball-mill device may induce the formation of organometallic compounds of a Grignard type [120, 121].

3. Concluding remarks

Understanding the mechanism of a hydrodechlorination process is important only if the degradation products are organic compounds of lesser toxicity. If the destruction is complete, yielding CO_2 and various carbonaceous structures, the mechanism is less important.

Upon the main two hydrodechlorination paths (radicalic versus ionic), it can be observed that both may be encountered, but there seem to be more evidence for the first, although in many cases the first step is the formation of radical anion through single-electron transfer. The type of activation (chemical, photochemical, thermal, cavitational, or mechanochemical) is instrumental in the type of reaction mechanism. Reaction conditions as well as reagents are also decisive in the type of mechanism the process will adopt. But through a better understanding of these aspects, a grasp on the reaction mechanism could be taken, and thus it could be possible to not only anticipate the advantages but also the limitations of a particular process. For example, when treating polluted soils with a mixture of Al and CaO, according to this type of mechanism, the presence of a hydrogen donor, namely soil moisture, is necessary [122]. Thus, it could be predicted that a soil too dry would be impracticable for treatment. On the other hand, many metallic systems can be effective even in the presence of large quantities of water [123], and even in aqueous media [124]. At the same time, the importance of the chemisorptions of hydrogen ions or nascent hydrogen on metallic surfaces has been understood, and the process can somehow be improved by the addition of stabilizers such as polyvinylpyrrolidone [125] or even biochar [126].

Thus, it will become easier to predict the successful outcome of a certain treatment process for a certain contaminated matrix (fly ash, soil, groundwater or wastewater) by considering the metallic system chosen, the moisture content (or the addition of any other potential hydrogen donor, in the form of an organic solvent—mainly alcohols), and the presence of a sorption substrate for the different forms of hydrogen transfer involved (even if it is only fly ash).

Author details

Yoshiharu Mitoma[1]*, Yumi Katayama[2] and Cristian Simion[3]

*Address all correspondence to: mitomay@pu-hiroshima.ac.jp

1 Department of Environmental Sciences, Prefectural University of Hiroshima, Shobara City, Hiroshima, Japan

2 Department of Life and Environmental Science, Hachinohe Institute of Technology, Hachinohe, Aomori Prefecture, Japan

3 Department of Organic Chemistry, Faculty of Applied Chemistry and Material Science, Politehnica University of Bucharest, Bucharest, Romania

References

[1] Carpenter DO. Polychlorinated biphenyls (PCBs): Routes of exposure and effects on human health. Reviews on Environmental Health. 2006;21(1):1-24

[2] Ni HG, Zeng H, Tao S, Zeng EY. Environmental and human exposure to persistent halogenated compounds derived from e-waste in China. Environmental Toxicology and Chemistry. 2010;29(6):1237-1247

[3] White SS, Birnbaum LS. An overview of the effects of dioxins and dioxin-like compounds on vertebrates, as documented in human and ecological epidemiology. Journal of Environmental Science and Health, Part C. 2009;27(4):197-211

[4] Annamalai J, Namasivayam V. Endocrine disrupting chemicals in the atmosphere: Their effects on humans and wildlife. Environment International. 2015;**76**:78-97

[5] Hens B, Hens L. Persistent threats by persistent pollutants: Chemical nature, concerns and future policy regarding PCBs—What are we heading for? Toxics. 2018;**6**:1-21

[6] Kulkarni PS, Crespo JG, Afonso CAM. Dioxins sources and current remediation technologies—A review. Environment International. 2008;**34**:139-153

[7] Gomes HI, Dias-Ferreira C, Ribeiro AB. Overview of in situ and ex situ remediation technologies for PCB-contaminated soils and sediments and obstacles for full-scale application. Science of the Total Environment. 2013;**445-446**:237-260

[8] Huang B, Lei C, Wei C, Zeng G. Chlorinated volatile organic compounds (Cl-VOCs) in environment – Sources, potential human health impacts, and current remediation technologies. Environment International. 2014;**71**:118-138

[9] Dai C, Zhou Y, Peng H, Huang S, Qin P, Zhang J, Yang Y, Luo L, Zhang X. Current progress in remediation of chlorinated volatile organic compounds: A review. Journal of Industrial and Engineering Chemistry. 2018;**62**:106-119

[10] Fan G, Wang Y, Fang G, Zhu X, Zhou D. Review of chemical and electrokinetic remediation of PCBs contaminated soils and sediments. Environmental Science: Processes and Impacts. 2016;**18**(9):1140-1156

[11] Cagnetta G, Robertson J, Huang J, Zhang K, Yu G. Mechanochemical destruction of halogenated organic pollutants: A critical review. Journal of Hazardous Materials. 2016;**313**:85-102

[12] Frascari D, Zanaroli G, Danko AS. In situ aerobic co-metabolism of chlorinated solvents: A review. Journal of Hazardous Materials. 2015;**283**:382-399

[13] Yang J, Meng L, Guo L. In situ remediation of chlorinated solvent-contaminated groundwater using ZVI/organic carbon amendment in China: Field pilot test and full-scale application. Environmental Science and Pollution Research. 2018;**25**(6):5051-5062

[14] Zhang M, Zhao D. In situ dechlorination in soil and groundwater using stabilized zero-valent iron nanoparticles: Some field experience on effectiveness and limitations in "Novel Solutions to Water Pollution", ACS Symposium Series, Vol. 1123, Chapter 6, pp. 79-96; 2013

[15] Sinkkonen S, Paasivirta J. Degradation half-life times of PCDDs, PCDFs and PCBs for environmental fate modeling. Chemosphere. 2000;**40**:943-949

[16] Poillon F. Dioxin Treatment Technologies. Congress of the United States, Office of Technology Assessment. Washington DC, USA: DIANE Publishing; 1991

[17] Zhang M, Buekens A. De novo synthesis of dioxins: A review. International Journal of Environment and Pollution. 2016;**60**(1):63-110

[18] Zhou H, Meng A, Long Y, Li Q, Zhang Y. A review of dioxin-related substances during municipal solid waste incineration. Waste Management. 2015;**36**:106-118

[19] Vallejo M, Fresnedo San Roman M, Ortiz I, Irabien A. Overview of the PCDD/Fs degradation potential and formation risk in the application of advanced oxidation processes (AOPs) to wastewater treatment. Chemosphere. 2015;**118**:44-56

[20] Tong M, Yuan S. Physiochemical technologies for HCB remediation and disposal: A review. Journal of Hazardous Materials. 2012;**229-230**:1-14

[21] Guemiza K, Coudert L, Metahni S, Mercier G, Besner S, Blais JF. Treatment technologies used for the removal of As, Cr, Cu, PCP and/or PCDD/F from contaminated soil: A review. Journal of Hazardous Materials. 2017;**333**:194-214

[22] Modak A, Maiti D. Metal catalyzed defunctionalization reactions. Organic and Biomolecular Chemistry. 2016;**14**:21-35

[23] Altarawneh M, Dlugogorski BZ, Kennedy EM, Mackie JC. Mechanisms for formation, chlorination, dechlorination and destruction of polychlorinated dibenzo-p-dioxins and dibenzofurans (PCDD/Fs). Progress in Energy and Combustion Science. 2009;**35**:245-274

[24] Bhatt P, Kumar MS, Mudliar S, Chakrabarti T. Biodegradation of chlorinated compounds—A review. Critical Reviews in Environmental Science and Technology. 2007;**37**(2): 165-198

[25] Bess EN, Bischoff AJ, Sigman MS. Designer substrate library for quantitative, predictive modeling of reaction performance. Proceedings of the National Academy of Science. 2014;**111**(41):14698-14703

[26] Addink R, Olie K. Mechanisms of formation and destruction of polychlorinated dibenzo-p-dioxins and dibenzofurans in heterogeneous systems. Environmental Science and Technology. 1995;**29**:1425-1435

[27] Schoonenboom MH, Zoetemeijer HE, Olie K. Dechlorination of octachlorodibenzo-p-dioxin and octachlorodibenzofuran on an alumina support. Applied Catalysis B-Environmental. 1995;**6**:11-20

[28] Huang H, Buekens A. On the mechanisms of dioxin formation in combustion processes. Chemosphere. 1995;**31**(9):4099-4117

[29] Addink R, Govers HAJ, Olie K. Isomer distribution of polychlorinated dibenzo-p-dioxins/dibenzofurans formed during de novo synthesis on incinerator fly ash. Environmental Science and Technology. 1998;**32**(13):1888-1893

[30] Altwicker ER, Milligan MS. Formation of dioxins: Competing rates between chemically similar precursors and de novo reactions. Chemosphere. 1993;**27**(1-3):301-307

[31] Altwicker ER. Relative rates of formation of polychlorinated dioxins and furans from precursor and *de novo* reactions. Chemosphere. 1996;**33**(10):1897-1904

[32] Milligan MS, Altwicker ER. Mechanistic aspects of the de novo synthesis of polychlo-rinated dibenzo-p-dioxins and furans in fly ash from experiments using isotopically labeled reagents. Environmental Science and Technology. 1995;**29**(5):1353-1358

[33] Ritter ER, Bozzelli JW. Pathways to chlorinated dibenzodioxins and dibenzofurans from partial oxidation of chlorinated aromatics by OH radical: Thermodynamics and kinetic insights. Combustion Science and Technology. 1994;**101**:153-169

[34] Tuppurainen K, Halonen I, Ruokojärvi P, Tarhanen J, Ruuskanen J. Formation of PCDDs and PCDFs in municipal waste incineration and its inhibition mechanisms: A review. Chemosphere. 1998;**36**(7):1493-1511

[35] Halonen I, Tuppurainen K, Ruuskanen J. Formation of aromatic chlorinated compounds catalyzed by copper and iron. Chemosphere. 1997;**34**(12):2649-2662

[36] Hinton WS, Lane AM. Characterisation of municipal solid waste incinerator fly ash promoting the formation of polychlorinated dioxins. Chemosphere. 1991;**22**(5-6):473-483

[37] Manninen H, Perkiö A, Vartiainen T, Ruuskanen J. Formation of PCDD/PCDF: Effect of fuel and fly ash composition on the formation of PCDD/PCDF in the co-combustion of refuse-derived and packaging-derived fuels. Environmental Science and Pollution Research. 1996;**3**(3):129-134

[38] Gullett B, Bruce K, Beach L, Drago A. Mechanistic steps in the production of PCDD and PCDF during waste combustion. Chemosphere. 1992;**25**(7-10):1387-1392

[39] Hinton WS, Lane AM. Synthesis of polychlorinated dioxins over MSW incinerator fly ash to identify catalytic species. Chemosphere. 1991;**23**(7):831-840

[40] Gullett B, Bruce K, Beach L. The effect of metal catalysts on the formation of polychlo-rinated dibenzo-p-dioxin and polychlorinated dibenzofuran precursors. Chemosphere. 1990;**20**(10-12):1945-1952

[41] Addink R, Schoonenboom MH. Metals as catalysts during the formation and decompo-sition of chlorinated dioxins and furans in incineration processes. Journal of the Air and Waste Management Association;**48**(1997):101-105

[42] Keane MA. Supported transition metal catalysts for hydrodechlorination reactions. ChemCatChem. 2011;**3**(5):800-821

[43] Ukisu Y, Iimura S, Uchida R. Catalytic dechlorination of polychlorinated biphenyls with carbon-supported noble metal catalysts under mild conditions. Chemosphere. 1996;**33**(8):1523-1530

[44] Ukisu Y, Kameoka S, Miyadera T. Catalytic dechlorination of aromatic chlorides with noble-metal catalysts under mild conditions: Approach to practical use. Applied Catalysis B: Environmental. 2000;**27**:97-104

[45] Ukisu Y, Miyadera T. Hydrogen-transfer hydrodechlorination of polychlorinated dibenzo-*p*-dioxins and dibenzofurans catalyzed by supported palladium catalysts. Applied Catalysis B: Environmental. 2003;**40**:141-149

[46] Ukisu Y, Miyadera T. Dechlorination of dioxins with supported palladium catalysts in 2-propanol solution. Applied Catalysis A: General. 2004;**271**:165-170

[47] Charman HB. Hydride transfer reactions catalysed by metal complexes. Journal of the Chemical Society B: Physical Organic. 1967;**0**:629-632

[48] Hashimoto Y, Uemichi Y, Ayame A. Low-temperature hydrodechlorination mechanism of chlorobenzenes over platinum-supported and palladium-supported alumina catalysts. Applied Catalysis A: General. 2005;**287**:89-97

[49] Choi H, Al Abed SR, Agarwal S. Catalytic role of palladium and relative reactivity of substituted chlorines during adsorption and treatment of PCBs on reactive activated carbon. Environmental Science and Technology. 2009;**43**:7510-7515

[50] Fueno H, Tanaka K, Sugawa S. Theoretical study of the dechlorination reaction pathways of octachlorodibenzo-p-dioxin. Chemosphere. 2002;**48**:771-778

[51] Nomiyama K, Tanizaki T, Ishibashi H, Arizono K, Shinohara R. Production mechanism of hydroxylated PCBs by oxidative degradation of selected PCBs using TiO_2 in water and estrogenic activity of their intermediates. Environmental Science and Technology. 2005;**39**:8762-8769

[52] Miao XS, Chu SG, Xu XB. Degradation pathways of PCBs upon UV irradiation in hexane. Chemosphere. 1999;**39**(10):1639-1650

[53] Chen JR, Kim D, Park JS, Gil KI, Yen TF. Reductive dechlorination of polychlorinated biphenyls (PCBs) by ultrasound-assisted chemical process (UACP). Environmental Earth Sciences. 2013;**69**:1025-1032

[54] Zhang F, Chen J, Zhang H, Ni Y, Liang X. The study on the dechlorination of OCDD with Pd/C catalyst in ethanol–water solution under mild conditions. Chemosphere. 2007;**68**:1716-1722

[55] Lu GN, Dang Z, Fennell DE, Huang W, Li Z, Liu CQ. Rules of thumb for assessing reductive dechlorination pathways of PCDDs in specific systems. Journal of Hazardous Materials. 2010;**177**:1145-1149

[56] Choi H, Al Abed SR. PCB congener sorption to carbonaceous sediment components: Macroscopic comparison and characterization of sorption kinetics and mechanism. Journal of Hazardous Materials. 2009;**165**:860-866

[57] Choi H, Agarwal S, Al Abed SR. Adsorption and simultaneous dechlorination of PCBs on GAC/Fe/Pd: Mechanistic aspects and reactive capping barrier concept. Environmental Science and Technology. 2009;**43**:488-493

[58] Sajiki H, Kume A, Hattori K, Hirota K. Mild and general procedure for Pd/C-catalyzed hydrodechlorination of aromatic chlorides. Tetrahedron Letters. 2002;**43**:7247-7250

[59] Izadifard M, Achari G, Langford CH. The pathway of dechlorination of PCB congener by a photochemical chain process in 2-propanol: The role of medium and quenching. Chemosphere. 2008;**73**:1328-1334

[60] Izadifard M, Langford CH, Achari G. Photocatalytic dechlorination of PCB 138 using leuco-methylene blue and visible light; reaction conditions and mechanisms. Journal of Hazardous Materials. 2010;**181**:393-398

[61] Matsunaga A, Yasuhara A. Dechlorination of PCBs by electrochemical reduction with aromatic radical anion as mediator. Chemosphere. 2005;**58**:897-904

[62] Noma Y, Mitsuhara Y, Matsuyama K, Sakai SI. Pathways and products of the degradation of PCBs by the sodium dispersion method. Chemosphere. 2007;**68**:871-879

[63] Miyoshi K, Nishio T, Yasuhara A, Morita M. Dechlorination of hexachlorobiphenyl by using potassium-sodium alloy. Chemosphere. 2000;**41**:819-824

[64] Cao L, Sun W, Zhang Y, Feng S, Dong J, Zhang Y, Rittmann BE. Competition for electrons between reductive dechlorination and denitrification. Frontiers of Environmental Science and Engineering. 2017;**11**(6):14

[65] Suresh S. Reductive remediation of pollutants using metals. The Open Waste Management Journal. 2009;**2**:6-16

[66] Wu BZ, Chen HY, Wang SJ, Wai CM, Liao W, Chiu KH. Reductive dechlorination for remediation of polychlorinated biphenyls. Chemosphere. 2012;**88**:757-768

[67] Shih YH, Hsu CY, Su YF. Reduction of hexachlorobenzene by nanoscale zero-valent iron: Kinetics, pH effect, and degradation mechanism. Separation and Purification Technology. 2011;**76**:268-274

[68] Varanasi P, Fullana A, Sidhu S. Remediation of PCB contaminated soils using iron nanoparticles. Chemosphere. 2007;**66**:1031-1038

[69] Kim JH, Tratnyek PG, Chang YS. Rapid dechlorination of polychlorinated dibenzo-p-dioxins by bimetallic and nanosized zerovalent iron. Environmental Science and Technology. 2008;**42**:4106-4112

[70] Fang Y, Al-Abed SR. Dechlorination kinetics of monochlorobiphenyls by Fe/Pd: Effects of solvent, temperature, and PCB concentration. Applied Catalysis B: Environmental. 2008;**78**:371-380

[71] Colombo A, Dragonetti C, Magni M, Roberto D. Degradation of toxic halogenated organic compounds by iron-containing mono-, bi- and tri-metallic particles in water. Inorganica Chimica Acta. 2015;**431**:48-60

[72] Zhou T, Li Y, Lim TT. Catalytic hydrodechlorination of chlorophenols by Pd/Fe nanoparticles: Comparisons with other bimetallic systems, kinetics and mechanism. Separation and Purification Technology. 2010;**76**:206-214

[73] O'Carroll D, Sleep B, Krol M, Boparai H, Kocur C. Nanoscale zero valent iron and bimetallic particles for contaminated site remediation. Advances in Water Resources. 2013;**51**:104-122

[74] Venkatachalam K, Arzuaga X, Chopra N, Gavalas VG, Xu J, Bhattacharyya D, Hennig B, Bachas LG. Reductive dechlorination of 3,30,4,40-tetrachlorobiphenyl (PCB77) using

palladium or palladium/iron nanoparticles and assessment of the reduction in toxic potency in vascular endothelial cells. Journal of Hazardous Materials. 2008;**159**:483-491

[75] Agarwal S, Al Abed SR, Dionysiou DD, Graybill E. Reactivity of substituted chlorines and ensuing dechlorination pathways of select PCB congeners with Pd/Mg bimetallics. Environmental Science and Technology. 2009;**43**:915-921

[76] Agarwal S, Al Abed SR, Dionysiou DD. Enhanced corrosion-based Pd/Mg bimetallic systems for dechlorination of PCBs. Environmental Science and Technology. 2007;**41**: 3772-3727

[77] Agarwal S, Al Abed SR, Dionysiou DD. Impact of organic solvents and common anions on 2-chlorobiphenyl dechlorination kinetics with Pd/Mg. Applied Catalysis B: Environmental. 2009;**92**:17-22

[78] Agarwal S, Al Abed SR, Dionysiou DD. A feasibility study on Pd/Mg application in historically contaminated sediments and PCB spiked substrates. Journal of Hazardous Materials. 2009;**172**:1156-1162

[79] Yang B, Deng S, Yu G, Zhang H, Wu J, Zhuo Q. Bimetallic Pd/Al particles for highly efficient hydrodechlorination of 2-chlorobiphenyl in acidic aqueous solution. Journal of Hazardous Materials. 2011;**189**:76-83

[80] Coutts JG, Devor RW, Aitken B, Hampton MD, Quinn JW, Clausen CA, Geiger CL. The use of mechanical alloying for the preparation of palladized magnesium bimetallic particles for the remediation of PCBs. Journal of Hazardous Materials. 2011;**192**:1380-1387

[81] DeVor R, Carvalho-Knighton K, Aitken B, Maloney P, Holland E, Talalaj L, Fidler R, Elsheimer S, Clausen CA, Geiger CL. Dechlorination comparison of mono-substituted PCBs with Mg/Pd in different solvent systems. Chemosphere. 2008;**73**:896-900

[82] DeVor R, Carvalho-Knighton K, Aitken B, Maloney P, Holland E, Talalaj L, Elsheimer S, Clausen CA, Geiger CL. Mechanism of the degradation of individual PCB congeners using mechanically alloyed Mg/Pd in methanol. Chemosphere. 2009;**76**:761-766

[83] Maloney P, DeVor R, Novaes-Card S, Saitta E, Quinn J, Clausen CA, Geiger CL. Dechlorination of polychlorinated biphenyls using magnesium and acidified alcohols. Journal of Hazardous Materials. 2011;**187**:235-240

[84] Mitoma Y, Katayama Y, Simion AM, Harada H, Kakeda M, Egashira N, Simion C. Considerations on on the mechanism of Ca/ethanol/Pd/C assisted hydrodechlorination of chlorinated aromatic substrates. Chemosphere. 2016;**164**:92-97

[85] Mitoma Y, Simion AM, Mallampati SR, Miyata H, Kakeda M, Simion C. Hydrodechlorination of PCDD/PCDF/PCB contaminants by simple grinding of contaminated soils with a nano-size calcium reagent. Environmental Progress and Sustainable Energy. 2016;**35**(1): 34-40

[86] Simion AM, Miyata H, Kakeda M, Egashira N, Mitoma Y, Simion C. Direct and complete cleansing of transformer oil contaminated by PCBs. Separation and Purification Technology. 2013;**103**:267-272

[87] Simion AM, Kakeda M, Egashira N, Mitoma Y, Simion C. A direct method for the decontamination of a fly ash amended wet soil, artificially polluted with dioxins. Central European Journal of Chemistry. 2012;10(5):1547-1555

[88] Mitoma Y, Kakeda M, Simion AM, Egashira N, Simion C. Metallic Ca–Rh/C-Methanol, a high-performing system for the hydrodechlorination/ring reduction of mono- and poly chlorinated aromatic substrates. Environmental Science and Technology. 2009;43(15):5952-5958

[89] Mitoma Y, Egashira N, Simion C. Highly effective degradation of polychlorinated biphenyls in soil mediated by a Ca/Rh bicatalytic system. Chemosphere. 2009;74:968-973

[90] Mitoma Y, Tasaka N, Takase M, Masuda T, Tashiro H, Egashira N. Calcium-promoted catalytic degradation of PCDDs, PCDFs, and coplanar PCBs under a mild wet process. Environmental Science and Technology. 2006;40:1849-1854

[91] Mitoma Y, Uda T, Egashira N, Simion C, Tashiro H, Tashiro M, Fan XB. Approach to highly efficient dechlorination of PCDDs, PCDFs, and coplanar PCBs using metallic calcium in ethanol under atmospheric pressure at room temperature. Environmental Science and Technology. 2004;38(4):1216-1220

[92] Ghaffar A, Tabata M. Dechlorination/detoxification of aromatic chlorides using fly ash under mild conditions. Waste Management. 2009;29:3004-3008

[93] Gao X, Wang W, Liu X. Dechlorination reaction of hexachlorobenzene with calcium oxide at 300-400°C. Journal of Hazardous Materials. 2009;169:279-284

[94] Su G, Huang L, Shi R, Liu Y, Lu H, Zhao Y, Yang F, Gao L, Zheng M. Thermal dechlorination of PCB-209 over Ca species-doped Fe_2O_3. Chemosphere. 2016;144:81-90

[95] Gao X, Wang W, Liu X. Low-temperature dechlorination of hexachlorobenzene on solid supports and the pathway hypothesis. Chemosphere. 2008;71:1093-1099

[96] Arulmozhiraja S, Morita M. Electron affinities and reductive dechlorination of toxic polychlorinated dibenzofurans: A density functional theory study. Journal of Physical Chemistry A. 2004;108:3499-3508

[97] Kim Y-H, Carraway ER. Dechlorination of chlorinated phenols by zero valent zinc. Environmental Technology. 2003;24(12):1455-1463

[98] Huang B, Qian W, Yu C, Wang T, Zeng G, Lei C. Effective catalytic hydrodechlorination of o-, p- and m-chloronitrobenzene over Ni/Fe nanoparticles: Effects of experimental parameter and molecule structure on the reduction kinetics and mechanisms. Chemical Engineering Journal. 2016;306:607-618

[99] Begum A, Gautam SK. Dechlorination of endocrine disrupting chemicals using Mg^0/ $ZnCl_2$ bimetallic system. Water Research. 2011;45:2383-2391

[100] Murena F, Schioppa E. Kinetic analysis of catalytic hydrodechlorination process of polychlorinated biphenyls (PCBs). Applied Catalysis B: Environmental. 2000;27:257-267

[101] Lim TT, Zhu BW. Practical applications of bimetallic nano-iron particles for reductive dehalogenation of haloorganics: Prospects and challenges. In: Carvalho-Knighton KM,

Geiger CL, editors. Environmental Applications of Nanoscale and Microscale Reactive Metal Particles. USA: American Chemical Society; 2009. pp. 245-261. (Chapter 14)

[102] Lin S, Su G, Zheng M, Jia M, Qi C, Li W. The degradation of 1,2,4-trichlorobenzene using synthesized Co$_3$O$_4$ and the hypothesized mechanism. Journal of Hazardous Materials. 2011;**192**:1697-1704

[103] Weber R, Yoshida S, Miwa K. PCB destruction in subcritical and supercritical water s evaluation of PCDF formation and initial steps of degradation mechanisms. Environmental Science and Technology. 2002;**36**:1839-1844

[104] Doornkamp C, Ponec V. The universal character of the Mars and Van Krevelen mechanism. Journal of Molecular Catalysis A: Chemical. 2000;**162**(1-2):19-32

[105] Zhang L, Zheng M, Liu W, Zhang B, Su G. A method for decomposition of hexachlorobenzene by γ-alumina. Journal of Hazardous Materials. 2008;**150**:831-834

[106] Khaleel A, Al-Nayli A. Supported and mixed oxide catalysts based on iron and titanium for the oxidative decomposition of chlorobenzene. Applied Catalysis B: Environmental. 2008;**80**:176-184

[107] King CM, King RB, Bhattacharyya NK, Newton MG. Organonickel chemistry in the catalytic hydrodechlorination of polychlorobiphenyls (PCBs): Ligand steric effects and molecular structure of reaction intermediates. Journal of Organometallic Chemistry. 2000;**600**:63-70

[108] Gorbunova TI, Subbotina JO, Saloutin VI, Chupakhin ON. Reactivity of polychlorinated biphenyls in nucleophilic and electrophilic substitutions. Journal of Hazardous Materials. 2014;**278**:491-499

[109] Fang G, Gao J, Dionysiou DD, Liu C, Zhou D. Activation of persulfate by quinones: Free radical reactions and implication for the degradation of PCBs. Environmental Science and Technology. 2013;**47**:4605-4611

[110] Zhang G, Hua I. Cavitation chemistry of polychlorinated biphenyls: Decomposition mechanisms and rates. Environmental Science and Technology. 2000;**34**:1529-1534

[111] Dubinskaya AM. Transformations of organic compounds under the actionof mechanical stress. Russian Chemical Reviews. 1999;**68**:637-652

[112] Birke V, Mattik J, Runne D. Mechanochemical reductive dehalogenation of hazardous polyhalogenated contaminants. Journal of Materials Science. 2004;**39**:5111-5116

[113] Ikoma T, Zhang Q, Saito F, Akiyama K, Tero-Kubota S, Kato T. Radicals in the mechanochemical dechlorination of hazardous organochlorine compounds using CaO nanoparticles. Bulletin of the Chemical Society of Japan. 2001;**74**:2303-2309

[114] Tanaka Y, Zhang Q, Saito F. Mechanochemical dechlorination of trichlorobenzene on oxide surfaces. Journal of Physical Chemistry B. 2003;**107**:11091-11097

[115] Zhang Q, Saito F, Ikoma T, Tero-Kubota S, Hatakeda K. Effects of quartz addition on the mechanochemical dechlorination of chlorobiphenyl by using CaO. Environmental Science and Technology. 2001;**35**:4933-4935

[116] Tanaka Y, Zhang Q, Saito F, Ikoma T, Tero-Kubota S. Dependence of mechanochemically induced decomposition of mono-chlorobiphenyl on the occurrence of radicals. Chemosphere. 2005;**60**:939-943

[117] Lu S, Huang J, Peng Z, Li X, Yan J. Ball milling 2,4,6-trichlorophenol with calcium oxide: Dechlorination experiment and mechanism considerations. Chemical Engineering Journal. 2012;**195-196**:62-68

[118] Zhang W, Wang H, Hiang J, Yu M, Wang F, Zhou L, Yu G. Acceleration and mechanistic studies of the mechanochemical dechlorination of HCB with iron powder and quartz sand. Chemical Engineering Journal. 2014;**239**:185-191

[119] Pizzigallo MDR, Napola A, Spagnuolo M, Ruggiero P. Mechanochemical removal of organo-chlorinated compounds by inorganic components of soil. Chemosphere. 2004;**55**:1485-1492

[120] Birke V, Schütt C, Ruck WKL. Small particle size magnesium in one-pot Grignard-Zerewitinoff-like reactions under mechanochemical conditions: On the kinetics of reductive dechlorination of persistent organic pollutants (POPs). In: Geiger CL, Carvalho-Knighton KM, editors. Environmental Applications of Nanoscale and Microscale Reactive Metal Particles. Washington DC: American Chemical Society; 2010. pp. 39-54

[121] Birke V, Schütt C, Burmeier H, Ruck WKL. Defined mechanochemical reductive dechlorination of 1,3,5-trichlorobenzene at room temperature in a ball mill. Fresenius Environmental Bulletin. 2011;**20**:2794-2805

[122] Jiang Y, Shang Y, Yu S, Liu J. Dechlorination of hexachlorobenzene in contaminated soils using a nanometallic Al/CaO dispersion mixture: Optimization through response surface methodology. International Journal of Environmental Research and Public Health. 2018;**15**:872

[123] Ghaffar A, Tabata M. Enhanced dechlorination of chlorobenzene compounds on fly ash: Effects of metals, solvents, and temperature. Green Chemistry Letters and Reviews. 2010;**3**(3):179-190

[124] Xia C, Xu J, W W, Liang X. Pd/C-catalyzed hydrodehalogenation of aromatic halides in aqueous solutions at room temperature under normal pressure. Catalysis Communications. 2004;**5**(8):383-386

[125] Fang L, Xu C, Zhang W, Huang LZ. The important role of polyvinylpyrrolidone and cu on enhancing dechlorination of 2,4-dichlorophenol by Cu/Fe nanoparticles: Performance and mechanism study. Applied Surface Science. 2018;**435**:55-64

[126] Zhu M, Zhang L, Zheng L, Zhuo Y, Xu J, He Y. Typical soil redox processes in pentachlorophenol polluted soil following biochar addition. Frontiers in Microbiology. 2018;**9**:579

Chapter 4

Application of Heterogeneous Catalysts in Dechlorination of Chlorophenols

Fuchong Li, Yansheng Liu, Linlei Wang, Xu Li,
Tianqiong Ma and Guangbi Gong

Additional information is available at the end of the chapter

http://dx.doi.org/10.5772/intechopen.79134

Abstract

Chlorophenols (CPs) is a very important kind of basic organic chemical intermediates such as sanitizers, germicides, insecticides and so on; but CPs also constitutes a particular group of priority pollutants that widely distribute in wastewater and the polluted groundwater. Because of their acute toxicity, persistence and low biodegradability, their emissions have been progressively restricted by strong legal regulations. In this chapter, we focused on methods for degrading of CPs recent years, especially by using new heterogeneous catalytic hydrogenation methods to the dechlorination of CPs. The purpose is to introduce scientific research workers and companies to waste water treatments in order to inspire and further better protect the environment.

Keywords: heterogeneous catalysts, dechlorination, chlorophenol, core-shell, environment

1. Introduction

The global water storage is up to 1.45 billion cubic kilometers (**Figure 1**), but the number of water storage can be produced and used directly by people is very less. Water is widely used in industries, agriculture, homes and so on, but keeping water quality becomes a major challenge for the coming decades. There are many factors that cause water damage. Industrial wastewater, agricultural wastewater, and domestic wastewater may be the main pollution sources (**Figure 2**).

Figure 1. Beautiful earth.

Figure 2. Chemical pollution is the largest source of pollution.

The development of chemical technology growing changed our lives. More and more food, medicine, cars, and appliance made by industries come to our lives. However, pollution problems were generating while we are enjoying our lives. Not only polluting the water and the air, but also seriously polluting the people's physical and health. Solving and treating the toxic organic compounds in wastewater are becoming an important project and research in recent years. It is becoming an important things for the government, companies, and academic research [1].

Chlorophenols (CPs) constitute a series of common organic compounds and intermediate involved in industrial processes such as fungicides, insecticides and dye precursors [2]. Chlorophenols (CPs) are also used in preservatives, papermaking and cosmetics industries. But because of acute toxicity, persistence and low biodegradability, chlorophenols especially such as chlorophenol, 2-chlorophenol, 2,4-dichlorophenol, 2,4,6-trichlorophenol, and 5-chlorophenol are highly toxic substances [3]. Their emissions have been progressively restricted by strong legal regulations. (Most chlorophenol compounds can interfere with endocrine, may cause cancer, cell mutations or teratogenesis, are almost non-biodegradable and difficult to remove from the environment. Therefore, they are a list of priority control pollutants by the US Environmental Protection Agency (EPA) and China Environmental Monitoring Center.)

In fact, chlorophenols and their derivatives may be produced by the chlorination of phenols. The phenol reacts with chlorination in the presence of high concentrations of chlorine and the toxicity of chlorophenols increases with its degree of chlorination and usually are difficult to degrade. These chlorophenols released into the natural environment can cause serious pollution, especially causing water pollution.

Conventional wastewater treatment based primarily on biological processes are not very efficient for the treatment of toxic or non-biodegradable and high-concentration wastewaters. Therefore, in order to protect our environment, it is necessary to develop efficient technologies for treating organic wastewater containing chlorine.

2. Application of heterogeneous catalysts in dechlorination

Recent years, the scientific community and the engineering technical community have begun to pay more attention to the CPs degradation or removal. In order to achieve this goal, large numbers of methods have been employed, such as oxidation (Fenton, O_3 and so on), aerobic/anaerobic biodegradation [4, 5], photocatalytic degradation [6], thermal combustion, catalytic reaction based on polymer membrane [7] and reduction dechlorination. In the chapter of this book, we classified these methods into two major categories: oxidation and reduction reactions in **Table 1**. In order to give the readers a clear understanding of the processing methods of the entire CPs treatments, we generally explain each method briefly. Because of heterogeneous catalysis is a cost-effective method for CPs treatments. We will give a detailed explanation and summary.

2.1. Photo-catalytic oxidation

Photocatalytic method [8, 9] is an advanced oxidation process (AOP). In common, semiconductor photocatalysts were needed to the oxidation reaction. Using higher than the semiconductor energy band gap light. For example, anatase which band gap is about 3.2 eV. Thus only ultraviolet light (wavelength <388 nm) can be effectively applied for electricity to be inspired to the conduction band from electronic valence band to generating a pair of holes and electrons in **Figure 3**. The hole–electron pair may be not stable enough. If they are not to combine together immediately, the two species (hole–electron pair) respectively may react with oxygen or water to generate oxidizing species, such as hydroxyl radicals, hydrogen oxygen free radical, superoxide free radicals and so on, hydroxyl radicals directly attack the phenol. The

Oxidation method				Reduction method
Photo-catalytic oxidation [8, 9]	Ultrasonic reaction	Fenton	O_3/CWAO Oxidation	Hydrodechlorination
Under light conditions, catalyst surface generate hole and electronics, then water reacts with oxygen free radicals.	Ultrasonic wave can greatly promote the AOP.	Fe^{2+} and H_2O_2 for generating hydroxyl radical (•OH).	•OH to degradation of organochlorine.	Via halogen atoms elimination with hydrogen.

Table 1. The classification of dechlorination reaction.

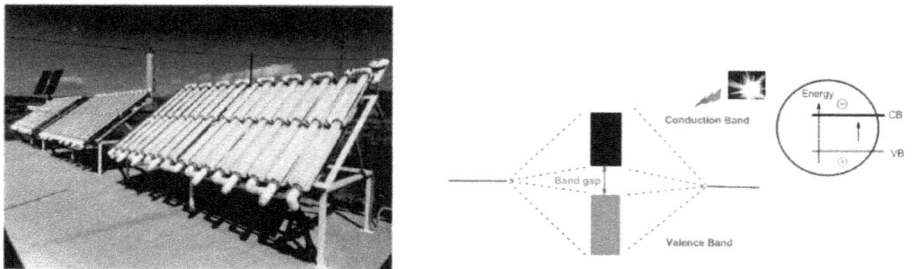

Figure 3. Theory of photocatalytic method.

proposed mechanism for the photocatalytic degradation of 4-chlorophenol by photocatalysts $(GR-Eu_2O_3/TiO_2)$ was described by Mallanaicker Myilsamy in **Figure 4** [10]:

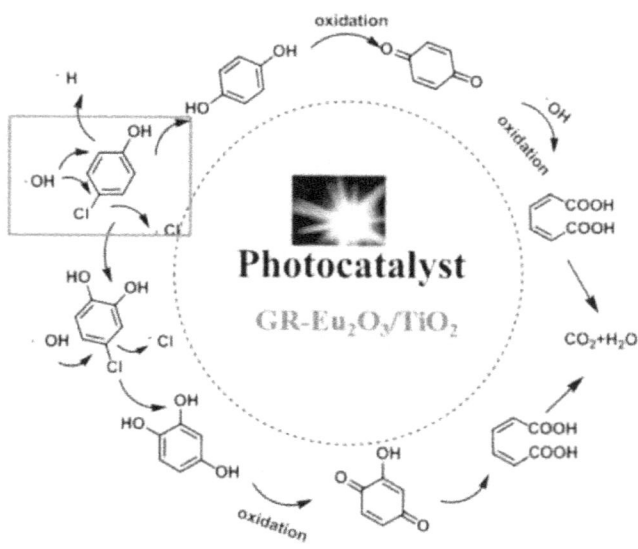

Figure 4. Proposed mechanism for the photocatalytic degradation of 4-chlorophenol [10].

There are several influence factors of this AOP.

The first: Photocatalysis: catalyst which determines the type of the light and speed of the reaction is a key factor we need to consider. Due to its high catalytic activity, low cost, chemical stability, non-toxicity, TiO_2 was usually applied as the most efficient photocatalyst in this AOP [11, 12]. Based on the photocatalytic mechanism of electron–hole, some metal nanoparticles were also used to prevent the combination of hole and electron pair [13, 14].

Other factors such as the amount of catalyst added, illumination angle, solvent, reaction temperature and so on will also affect the whole reaction efficiency [9].

Recent research directions focused on the development and synthesis of improved photocatalytic materials (such as core-shell nanoparticles, bimetals catalyst and so on), with two major lines aiming at (i) better charge separation and better charge migration (ii) visible-light driven photocatalysis. Claude Descorme has summed up most of the catalyst types in his review [9].

2.2. Sonocatalysis (ultrasonic reaction method)

Ultrasonic reaction (sonocatalysis) is also considered as an advanced oxidation process (AOP). It is well known that ultrasonic wave can greatly promote heat movement of molecules in the reaction, accelerating the reaction process, thus lots of scientists like to choose ultrasonic reaction method to the low reaction activities reagents, the entire process reaction temperature is not too high. The chemical processes of the sonocatalysis may involve two parts (I) free radical process and (II) pyrolysis process under ultrasonic wave. Recent research directions focused on the development of the more efficient reactor, for example (**Figure 5**), and catalyst which can be applied to decrease the frequencies [15, 16].

2.3. Fenton method

Fenton method [17] is also a AOP method which uses ferrous or cuprous salts ($Fe^{2+} + Cu^{2+}$) and hydrogen peroxide (H_2O_2) for generating powerful oxidant, i.e., hydroxyl radicals produced in situ **Figure 6**.

For the Fenton method to the dechlorination of Chlorophenols, the important parameter is probably the operating pH, In the presence of a homogeneous catalyst, an optimum is generally observed around pH 3. In fact, when pH < 2.5, the Fe(II) $(H_2O)_2+$ will generate while the pH higher than 4, free iron and reduced concentration of hydroxyl precipitation of iron oxide. The pH can also affect the stability of the heterogeneous catalyst, now most research has focused on the development of more stable and regeneration or be used in a wide pH range catalyst [18].

2.4. O_3 oxidation

Ozone oxidation is also a kind of VOP, both ozone oxidation and Fenton oxidation also produce free radicals such as •OH to degradation of organochlorine, especially for the organochlorine

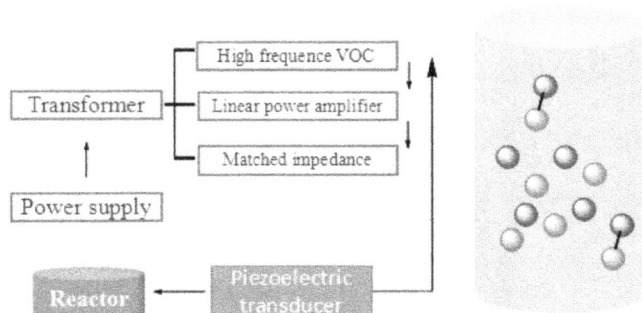

Figure 5. Scheme of the self-excited ultrasonic irradiation device.

$$Fe^{2+} + H_2O_2 / Fe^{3+} + OH^- + HO^· \qquad (1)$$
$$OH^- + RX / RX^{-·} + OH^- \qquad (2)$$
$$RH + OH^- / Rc + H_2O \qquad (3)$$
$$RHX + OH^- / RHX(OH) \qquad (4)$$

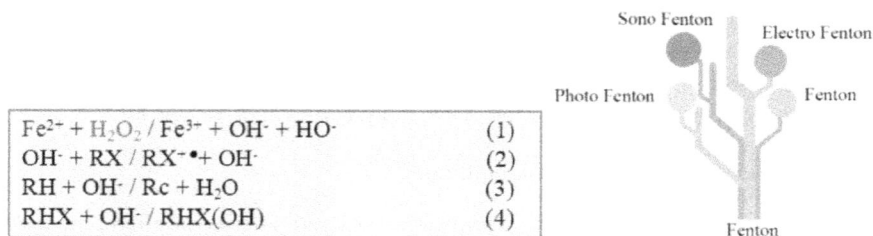

Figure 6. The mechanism of Fenton method.

pollution which contain multiple bonds such as $C = N$, $C = C$ has a better effect; while some limitations still exist in terms of the O_3 oxidation, chlorophenol concentrations, catalyst costs and its recovery and so on.

2.5. Hydrodechlorination

Heterogeneous catalyst replaced the homogeneous catalyst because of solving the catalyst recycling problem, especially for the precious metal catalysts, the catalyst costs were reduced. Heterogeneous catalyst have some advantages (**Figure 7**) compared with homogeneous catalyst, e.g., (I) Large surface area, especially for the porous heterogeneous catalyst, it may increase the reactant transmission rate in the catalyst. For gas phase reaction, surface adsorption effect will greatly increase the rate of the reaction. (II) The activity site will be discrete well in order to prevent its aggregation. (III) Diversified structure monomers to build diversified structure, morphology and functional catalyst. Wang Wei et al. [19], detailed introduced the nanocatalyst materials development which can be divided into three categories: inorganic materials, organic materials, and hybrid materials (**Figure 8**) and depicts some representative examples ranging from 0D discrete materials to 3D extended structures, from inorganic to pure organic components, from disordered to regular arrangements, and from non-porous to porous nature. In recent years, covalent organic frameworks attract more and more people's attention because of its large surface area and regular structure [20–22].

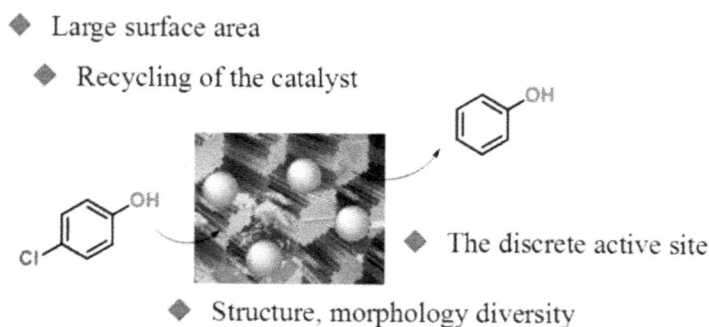

Figure 7. Heterogeneous catalysts applied in HDC.

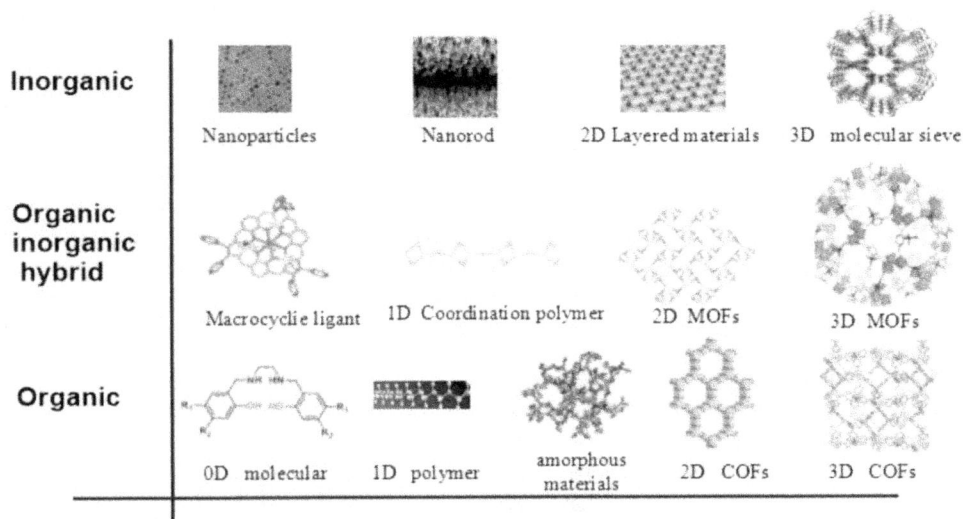

Figure 8. The classification of heterogeneous catalyst carrier and its development [19].

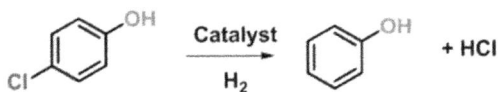

Figure 9. Hydrogenated dechlorination (HDC).

From 0D to 3D nanomaterials, all of them may be likely to be good candidates as the heterogeneous catalyst. Organic chlorides, especially of 4-chloride-phenol is a kind of highly toxic compound. Hydrogenated dechlorination (HDC) is an efficient method for toxic compound treating. Although this method does not lead to complete degradation of the organic chlorides pollutant but generates a less toxic (biodegradable) compound or a chemical of possible commercial important product via halogen atoms elimination with hydrogen (**Figure 9**) hydrochloric acid was produced in HDC reaction process in stoichiometric amount and the catalyst deactivation in this acid conditions is an important issue.

Recently most research focused on the development of more stable and efficient catalyst designed. According to the type and number of the metal, catalyst can be divided into single and bimetals center catalyst activity. Hence, a rapid, efficient, and green approach to the fabrication of effective catalyst Al_2O_3 shows a highly mechanical resistance with high catalytic activity and the potential applications the HDC of 4-CP has significant meaning for environmental protection. Pd^0 as the most effective metal active center for the HDC reaction usually applied with other metal cooperate to improve the catalytic efficiency and stability, like Pd-Bi [23, 24], Pd–Tl [25], Pd-Fe [26], and Ni@Pd [27] also have been extensively researched, in addition, the catalyst carrier has become the important role in improving the activity and stability of catalytic, such as Pd/Al_2O_3, Pd/zeolites, Pd/activated carbon (AC), SiO_2 supported Pd nanocatalysts.

As we know, for organic chloride, with chlorine content is higher, the toxicity become stronger at the same time, to be more difficult to removal chlorine from the organic chloride compound. Bimetal catalyst was reported to solve this issue by Anwar [28]. The pentachlorophenol (PCP, 10–20 mg min^{-1}) through the bimetal catalyst Ag^0/Fe^0 loaded on the heating column (25 × 1 cm) in scCO$_2$(supercritical carbon dioxide) was removed. After the reaction was operated for 1 h at 450°C, organically-bound chlorine was liberated, virtually quantitatively, from a 20% (w/v) feedstock stream (0.1 ml min^{-1} merged with 4 ml min^{-1} scCO$_2$). In extended operations, about 70°C, the reactor to 14 h of continuous running, no obvious activity loss if the chloride ion was washed every 3 h. In fact, catalytic activity was also related with the reaction temperature, pressure, solvent, and so on. The production of the pentachlorophenol (PCP) was a series of methylated phenols or methylated benzenes (**Figure 10**).

Jovanovic studied the dechlorination of p-chlorophenol in a microreactor with bimetallic Pd/Fe catalysts [29]. The bimetallic catalyst Pd/Fe was prepared by electroless deposition of Pd on the reactor plate surfaces. The chemistry of the dechlorination of p-chlorophenol on Pd/Fe catalysts involves three sets of chemical reactions (**Figure 11**) namely surface reactions, solution reactions, and actual dechlorination reactions. These reactions are found to be dependent on several parameters, including the pH, the Pd/Fe interface area, the extent of palladization, the ratio of the Pd/Fe interfacial area to the amount of chlorine to be removed, and the amount of dissolved oxygen.

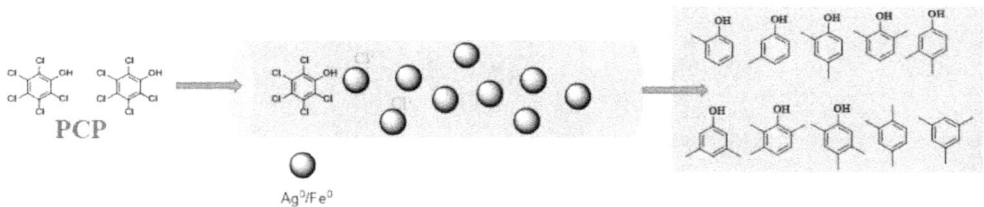

Figure 10. Pentachlorophenol treated by Ag^0/Fe^0 catalyst [28].

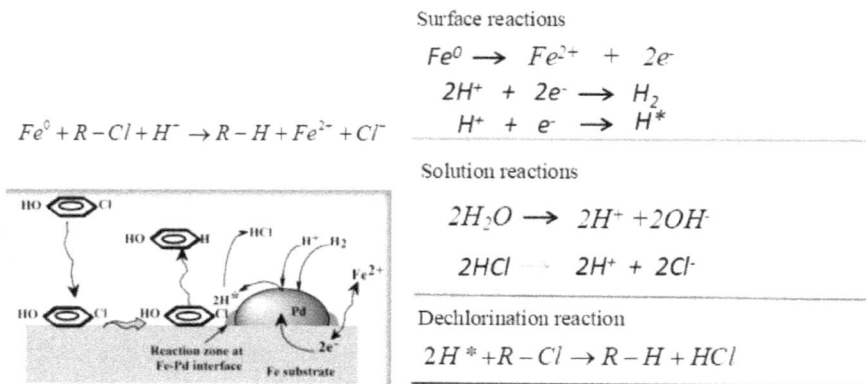

Figure 11. Dechlorination of p-chlorophenol on Pd/Fe catalysts.

2.5.1. How to design and synthesize catalyst of the HDC

- Catalyst carrier with large surface area to increase the adsorption quantity of H_2 in order to increase the speed of the reaction rate while could reduce the pressure.

- Improve the recovery efficiency of nanocatalyst.

- Prevent the by-products hydrochloric acid damage to the catalyst.

2.5.2. Core-shell catalyst

Nanometer catalysts because of their nanosize effect may be an efficient heterogeneous catalyst model which are widely used in many different catalytic reaction, such as, catalysis, biological

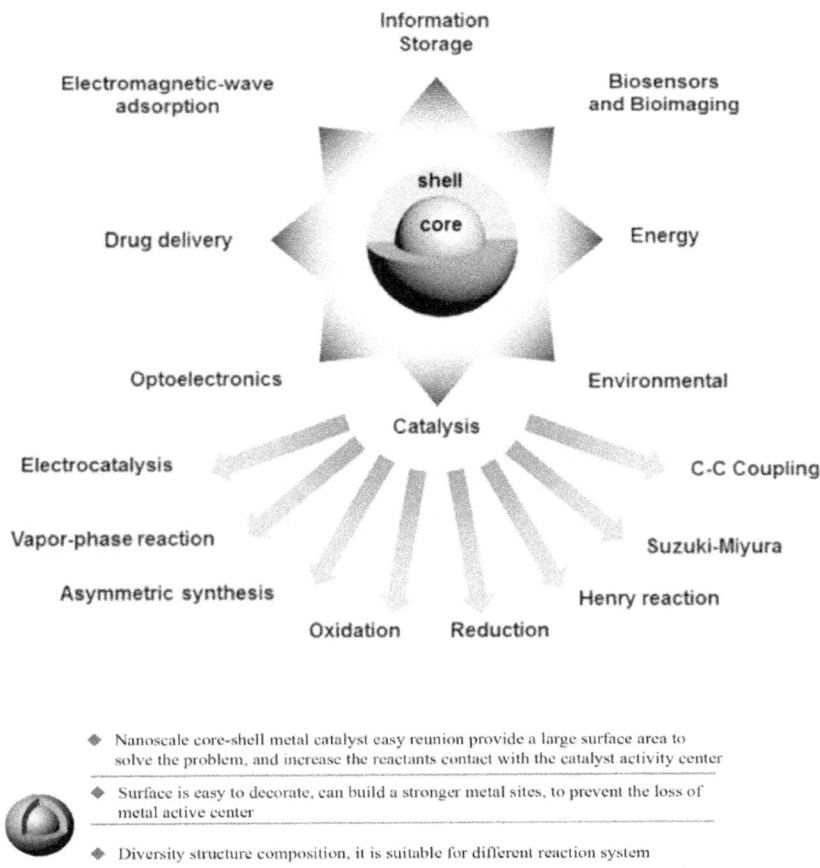

Figure 12. The advantages of core-shell nanoparticles.

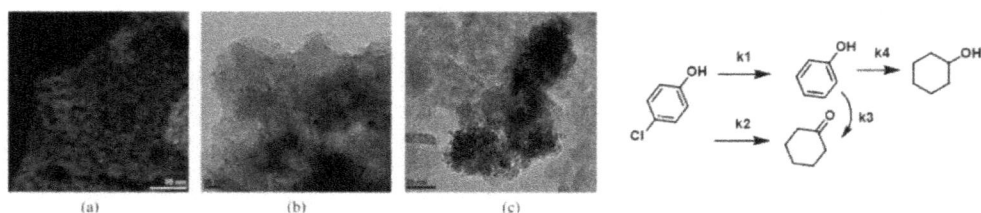

Figure 13. STEM micrograph of Pd/Al$_2$O$_3$ catalyst (a) and TEM micrographs of Pd/AC (b) and Rh/AC catalysts(c).

T (°C)	Kinetic constant (L/kg$_{cat}$ h)	Pd/Al$_2$O$_3$		Pd/AC		Rh/AC	
		k	r^2	k	r^2	k	r^2
20	k$_1$	146 ± 5	0.995	179 ± 36	0.991	44 ± 5	0.992
	k$_2$	10 ± 5		0		18 ± 5	
	k$_3$	9 ± 3		10 ± 4		7 ± 5	
30	k$_1$	201 ± 11	0.976	460 ± 37	0.976	96 ± 8	0.986
	k$_2$	29 ± 10		0		41 ± 8	
	k$_3$	20 ± 5		22 ± 7		13 ± 8	
	k$_4$	—		—		8 ± 4	
40	k$_1$	478 ± 16	0.994	713 ± 36	0.992	172 ± 13	0.982
	k$_2$	66 ± 12		113 ± 19		65 ± 12	
	k$_3$	45 ± 4		10 ± 3		11 ± 6	
	k$_4$	—		—		10 ± 4	

Table 2. The kinetics of hydrodechlorination of 4-chlorophenol on Pd/Al$_2$O$_3$, Pd/AC, Rh/AC.

medicine, material chemistry, sensors and so on [30]. As a nanomaterials, core-shell structure catalyst materials can be synthesized through the grafting reaction method step by step.

Because of the advantages that core-shell nanomaterials own (**Figure 12**) precious metals (Pd, Rh, Au…) can be loaded on core-shell nanoparticles by stronger chemical bonds, such as coordination bond or covalent bonds.

Elena Diaz et al. [31] studied the kinetics of hydrodechlorination of 4-chlorophenol on alumina and activated carbon supported Pd and Rh catalysts. The hydrodechlorination of 4-chlorophenols based on Pd and Rh on γ-alumina and activated carbon was investigated in continuously stirred basket reactors (20–40°C and 1 bar). For 4-chlorophenol, the reaction rate shows a first-order dependence. All catalysts are effective in removing 4-chlorophenol. Phenol, cyclohexanone and cyclohexanol were identified as reaction products (**Figure 13**). The hydrogenation of 4-chlorophenol to phenol was in the range of 146–478 L/kgcat h for Pd/Al$_2$O$_3$, 179–713 L/kgcat h for Pd/AC, and 44–172 L/kgcat h for Rh/AC. In all cases, the k$_1$

constant shows a much larger value than k_2, indicating that the formation of phenol is superior to cyclohexanone as the first reaction step (**Table 2**).

Wu et al. [32] designed and synthesized $Fe_3O_4@SiO_2@Pd$-Au catalyst (**Figure 14**). $Fe_3O_4@SiO_2@$ Pd-Au was synthesized with the reduction of Pd^{2+} and Au^{3+}. The amine-modified silica is coated on the outer layer of magnetic Fe_3O_4 nanoparticles to be a carrier for Pd-Au nanoparticles, where in the amine acts as a bridge connecting the Pd-Au nanoparticles to the support, making it highly dispersible, the magnetic properties of Fe_3O_4. allows the catalyst to be recycled. The performance of the catalyst was evaluated by hydrodechlorination of 4-chlorophenol (25°C, atmospheric pressure, a certain amount of 4-CP, 0.05 g NaOH, 0.5 g catalyst). The results showed that Pd nanoparticles have higher activity in HDC of 4-CP then the Au

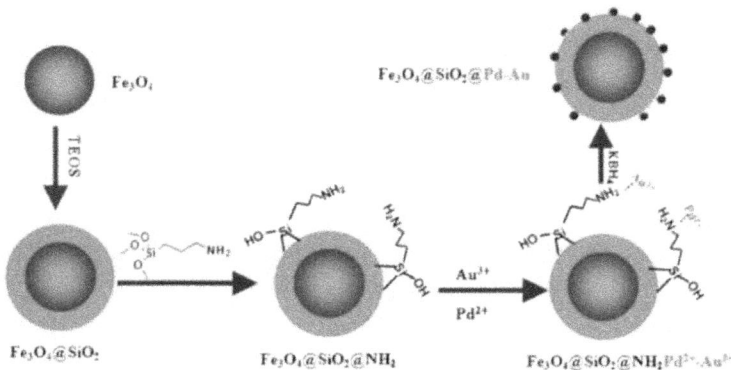

Figure 14. Synthesis procedure of $Fe_3O_4@SiO_2@Pd$-Au catalyst.

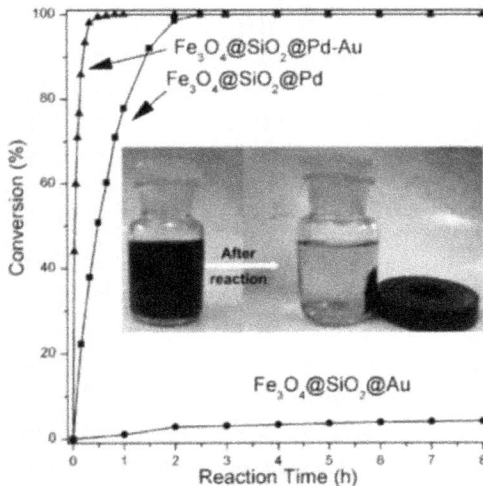

Figure 15. 4-CP HDC conversions on different catalysts.

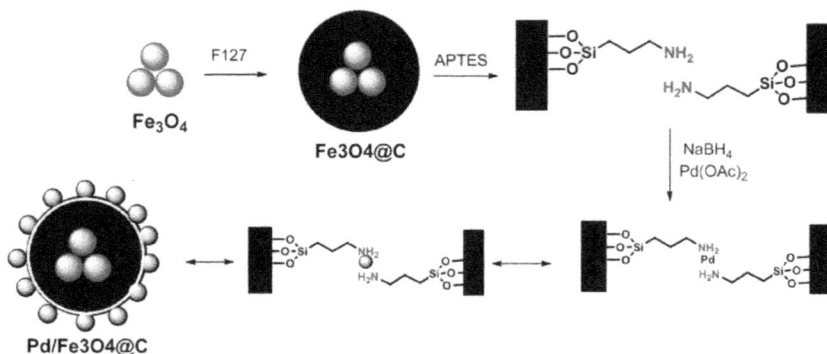

Figure 16. Design and synthesize the Pd/Fe$_3$O$_4$ @C catalyst.

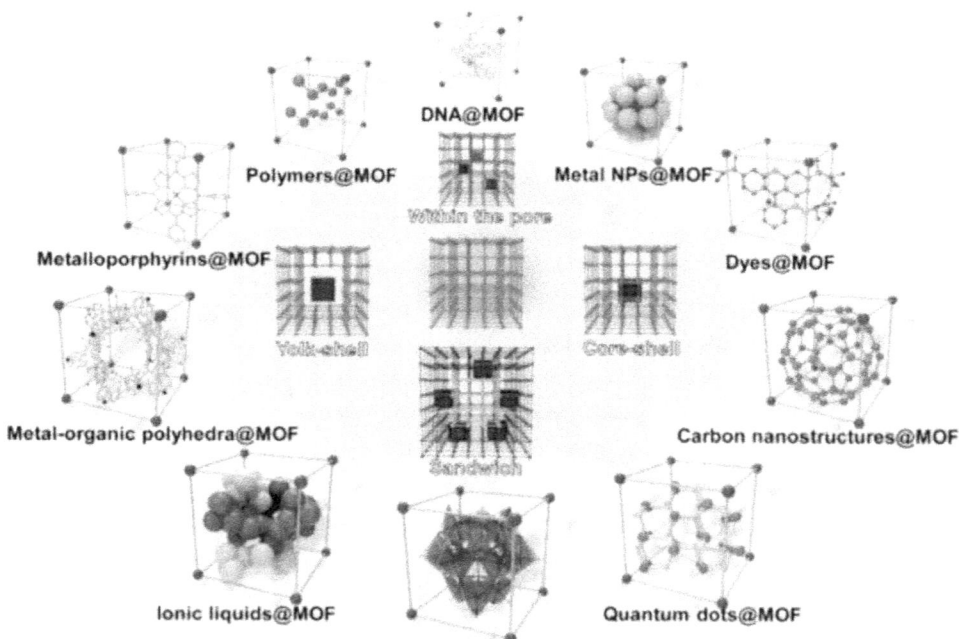

Figure 17. Controllable design of tunable nanostructures inside metal–organic frameworks.

nanoparticle, the Pd-Au alloy increases the conversion significantly and achieves a complete conversion of HDC within 20 min, much faster than the Pd metal catalyst (**Figure 15**).

Li et al. [33]., studies a high efficient Pd nanocatalysts (Pd/Fe$_3$O$_4$@C) applied in HDC. Pd nanoparticles in magnetic carbon shell can effectively improve the catalytic activity, separation and reusable. Catalyst was synthesized by Fe$_3$O$_4$ nanoparticles as a core, and then a layer of carbon layer on the outside package, finally by APTES modified carbon layer improving Pd

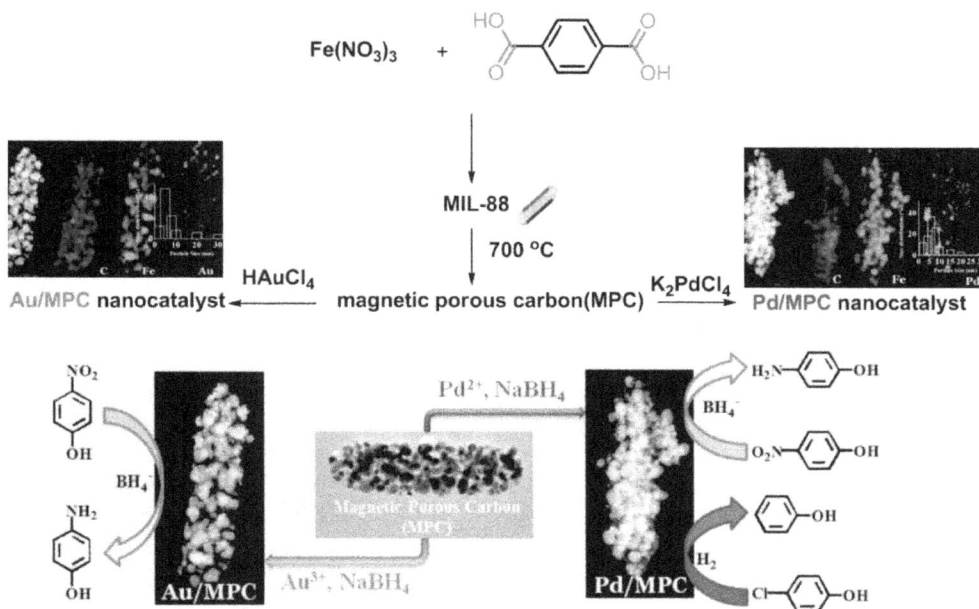

Figure 18. MPC utilized as a catalyst support to fabricate Au and Pd NP-based nanocatalysts.

loaded (**Figure 16**). $Pd/Fe_3O_4@C$ can be recycled at least five times without obvious loss activity. $Pd/Fe_3O_4@C$ is not only used in the aqueous solution of 4-chlorophenol hydrogenation dechlorination but also be used for reduction of 4-nitrophenol.

Porous materials are widely used in catalysis, gas adsorption, separation and other fields because of their large surface area and structure controllable. Metal organic frameworks (MOFs) materials are a kind of regular porous material formed by DCC chemistry by metal and organic ligand [34]. Controllable design of adjustable nanostructures in metal-organic frameworks is in **Figure 17**.

Noble metal nanoparticles (NMNPs) have attracted attention as the activity center. But NMNPs were easy blocking pores of active carbon, and leaching from carbon nanotubes, and graphene faces in the process of catalytic. In order to solve these problems, Dong used metal organic framework (MOF) to synthesize the porous carbon (MPC) which can provides a large surface area and pore (**Figure 18**) not only can make the active center (Pd NPs) scattered on it well, but also has paramagnetic behavior that the catalyst can be easily recycled [35].

3. Conclusion and outlook

Hydrogenated dechlorination degradation is an important method for the organochlorine degradation. Catalyst activity and its lifetime are two key points for the HCP, especially for the noble metal catalyst. In order to solve the recycling or reunion problems of noble metal catalyst, designing a suitable catalyst support becomes more important, core-shell type, organic porous materials, especially organic crystal type of porous materials, which is developing rapidly in recent years, such as MOF (metal organic frameworks) and COFs (covalent organic frameworks) materials because of the rules of uniform structure and large specific surface area that can bring some advantages to metal catalyst loading, design and synthesis coordination bond or other stronger bond to combine the metal and carrier is one of the important measure to prevent the catalyst loss.

Author details

Fuchong Li[1], Yansheng Liu[2], Linlei Wang[1], Xu Li[1], Tianqiong Ma[2] and Guangbi Gong[1]*

*Address all correspondence to: gongguangbi@petrochina.com.cn

1 Lanzhou Petrochemical Research Center, Lanzhou, PR China

2 Gansu Provincial Engineering Laboratory for Chemical Catalysis, College of Chemistry and Chemical Engineering, Lanzhou University, Lanzhou, PR China

References

[1] Pera-Titus M, Garcia-Molina V, Banos MA. Degradation of chlorophenols by means of advanced oxidation processes: A general review. Applied Catalysis B: Environmental. 2004;47:219-256. DOI: 10.1016/j.apcatb.2003.09.010

[2] Ramamoorthy S, Ramamoorthy S. Chlorinated Organic Compounds in the Environment: Regulatory and Monitoring Assessment. Vol. 9. Lewis Publishers; 1997. p. 1538. ISBN: 1566700418

[3] de Pedro ZM, Diaz E, Mohedano AF, Casas JA, Rodriguez JJ. Compared activity and stability of Pd/Al$_2$O$_3$ and Pd/AC catalysts in 4-chlorophenol hydrodechlorination in different pH media. Applied Catalysis B: Environmental. 2011;103:128-135. DOI: 10.1016/j.apceatb.2011.01.018

[4] Buitron G, Schoeb ME, Moreno-Andrade L, Moreno JA. Evaluation of two control strategies for a sequencing batch reactor degrading high concentration peaks of 4-chlorophenol. Water Research. 2005;39:1015-1024. DOI: 10.1016/j.watres.2004.12.023

[5] Majumder PS, Gupta SK. Degradation of 4-chlorophenol in UASB reactor under methanogenic conditions. Bioresource Technology. 2008;99:4169-4177. DOI: 10.1016/j.biortech.2007.08.062

[6] Ormad MP, Ovelleiro JL, Kiwi J. Photocatalytic degradation of concentrated solutions of 2,4-dichlorophenol using low energy light identification of intermediates. Applied Catalysis B: Environmental. 2001;(3):157-166

[7] Fritsch D, Kuhr K, Mackenzie K, Kopinke FD. Hydrodechlorination of chlorooganic in ground water by palladium catalysts part 1. Development of polymer-based catalysts and membrane reactor tests. Catalysis Today. 2003;82:105-118. DOI: 10.1016/S0920-5861(03)00208-6

[8] Ge T, Han J, Qi Y. The toxic effects of chlorophenols and associated mechanisms in fish. Aquatic Toxicology. 2017;**184**:78-93. DOI: 10.1016/j.aquatox.2017.01.005

[9] Claude Descorme. Catalytic wastewater treatment: Oxidation and reduction processes. Recent studies on chlorophenols. Catalysis Today. 2017;**297**:324-334. DOI: org/10.1016/j.cattod.2017.03.039

[10] Myilsamy M, Mahalakshmi M, Subha N. Visible light responsive mesoporous graphene-Eu_2O_3/TiO_2 nanocomposites for the efficient photocatalytic degradation of 4-chlorophenol. RSC Advances. 2016;**6**:35024-35035. DOI: 10.1039/C5RA27541E

[11] Zhao W, Ma W, Chen C. Efficient degradation of toxic organic pollutants with $Ni_2O_3/TiO_{2-x}B_x$ under visible irradiation. Journal of the American Chemical Society. 2004;**126**(15):4782-4783. DOI: 10.1021/ja0396753

[12] Kim W, Tachikawa T. Efficient degradation of toxic organic pollutants with $Ni_2O_3/TiO_{2-x}B_x$ under visible irradiation. Journal of Physical Chemistry C. 2009;**113**(24):10603-10609. DOI: 10.1021/ja0396753

[13] Zhu Z, Liu F, Zhang H. Photocatalytic degradation of 4-chlorophenol over Ag/MFe_2O_4 (M = Co, Zn, Cu, and Ni) prepared by a modified chemical co-precipitation method: A comparative study. RSC Advances. 2015;**5**:55499-55512. DOI: 10.1039/C5RA04608D

[14] Huerta Aguilar CA, Pandiyan T, Arenas-Alatorre JA, Singh N. Oxidation of phenols by TiO_2-Fe_3O_4-M (M = Ag or Au) hybrid composites under visible light. Separation and Purification Technology. 2015;**149**:265-278. DOI: org/10.1016/j.seppur.2015.05.019

[15] Papadaki M, Emery RJ, Abu-Hassan MA, Dıaz-Bustos A, Metcalfe IS, Mantzavinos D. Sonocatalytic oxidation processes for the removal of contaminants containing aromatic rings from aqueous effluents. Separation and Purification Technology. 2004;**34**: 35-42. DOI: 10.1016/S1383-5866(03)00172-2

[16] Dai Z, Chen A, Kisch H. Efficient Sonochemical Degradation of 4-Chlorophenol Catalyzed by Titanium Dioxide Hydrate. Chemistry Letters. 2005;**34**(12):1706. DOI: 10.1246/cl.2005.1706

[17] Posada D, Betancourt P, Liendo F. Catalytic wet air oxidation of aqueous solutions of substituted phenols. Catalysis Letters. 2006;**106**:81-88. DOI: 10.1007/s10562-005-9195-2

[18] Subramanian G, Madras G. Supplementary information remarkable enhancement of Fenton degradation at wide pH range promoted by thioglycolic acid. Chemical Communications. 2017;**53**:1136-1139. DOI: 10.1039/C6CC09962A

[19] Ding S-Y, Wang W. Covalent organic frameworks (COFs): From design to applications. Chemical Society Reviews. 2013;**42**:548-568. DOI: 10.1039/c2cs35072f

[20] Côté AP, Benin AI, Ockwig NW. Porous, crystalline, covalent organic frameworks. Science. 2005;**310**:1166-1170. DOI: 10.1126/science.112041

[21] Thomas A. Functional materials: From hard to soft porous frameworks. Angewandte Chemie, International Edition. 2010;**49**:8328-8344. DOI: 10.1002/anie.201000167

[22] Ding S-Y, Gao J, Wang Q. Construction of covalent organic framework for catalysis: Pd/COF-LZU1 in Suzuki–Miyaura coupling reaction. American Chemical Society. 2011;**133**:19816-19822. DOI: 10.1021/ja206846p

[23] Witońska I, Królak A, Karski S. Bi modified Pd/support (SiO_2, Al_2O_3) catalysts for hydrodechlorination of 2,4-dichlorophenol. Journal of Molecular Catalysis A: Chemical. 2010; **331**:21-28. DOI: org/10.1016/j.molcata.2010.07.011

[24] Karski S. Activity and selectivity of Pd–Bi/SiO_2 catalysts in the light of mutual interaction between Pd and Bi. Journal of Molecular Catalysis A: Chemical. 2006;**253**:147-154. DOI: org/10.1016/j.molcata.2006.03.013

[25] Karski S, Witońska I, Gołuchowska J. Catalytic properties of Pd–Tl/SiO_2 systems in the reaction of liquid phase oxidation of aldoses. Journal of Molecular Catalysis A: Chemical. 2006;**245**:225-230. DOI: org/10.1016/j.molcata.2005.10.004

[26] Izabela AW, Michael JW, Binczarski M . Pd–Fe/SiO_2 and Pd–Fe/Al_2O_3 catalysts for selective hydrodechlorination of 2,4-dichlorophenol into phenol. Journal of Molecular Catalysis A: Chemical. 2014;**393**:248-256. DOI: org/10.1016/j.molcata.2014.06.022

[27] Dong Z, Le X, Dong C. Ni@Pd core–shell nanoparticles modified fibrous silica nano spheres as highly efficient and recoverable catalyst for reduction of 4-nitrophenol and hydrodechlorination of 4-chlorophenol. Applied Catalysis B: Environmental. 2015;**162**: 372-380. DOI: org/10.1016/j.apcatb.2014.07.009

[28] Kabir A, Marshall WD. Dechlorination of pentachlorophenol in supercritical carbon dioxide with a zero-valent silver–iron bimetallic mixture. Green Chemistry. 2001;**3**:47-51. DOI: 10.1039/b008122l

[29] Goran N. Jovanovic, Polona Zÿ nidaršič Plazl, Ploenpun Sakrittichai, Khaled Al-Khaldi. Dechlorination of p-Chlorophenol in a Microreactor with Bimetallic Pd/Fe Catalyst. Industrial and Engineering Chemistry Research. 2005;**44**:5099-5106. DOI: 10.1021/ie049496+

[30] Gawande MB, Goswami A, Asefa T. Core–shell nanoparticles: Synthesis and applications in catalysis and electrocatalysis. Chemical Society Reviews. 2015;**44**:7540-7590. DOI: 10.1039/C5CS00343A

[31] Elena D, Casas JA, Mohedano ÁF. Kinetics of 4-chlorophenol hydrodechlorination with alumina and activated carbon-supported Pd and Rh catalysts. Industrial and Engineering Chemistry Research. 2009;**48**:3351-3358. DOI: 10.1021/ie801462b CCC: $40.75

[32] Zhijie W, Sun C, Chai Y. Fe_3O_4@SiO_2@Pd-Au: A highly efficient and magnetically separable catalyst for liquid-phase hydrodechlorination of 4-chlorophenol. RSC Advances. 2011;**1**:1179-1182. DOI: 10.1039/c1ra00491c

[33] Chen C, Li Y, Jia X. Wavelength-focusing organic molecular materials with diazoacetate or fumarate as a monofluorophore. New Journal of Chemistry. 2017;**41**:3693-3709. DOI: 10.1039/C7NJ90039B

[34] Chen L, Luque R, Li Y. Controllable design of tunable nanostructures inside metal–organic frameworks. Chemical Society Reviews. 2017;**46**:4614-4630. DOI: 10.1039/c6cs00537c

[35] Dong Z, Le X, Liu Y. Metal organic framework derived magnetic porous carbon composite supported gold and palladium nanoparticles as highly efficient and recyclable catalysts for reduction of 4-nitrophenol and hydrodechlorination of 4-chlorophenol. Journal of Materials Chemistry A. 2014;**2**:18775-18785. DOI: 10.1039/C4TA04010D

www.ingramcontent.com/pod-product-compliance
Lightning Source LLC
Chambersburg PA
CBHW081244190326
41458CB00016B/5916